METHODS OF
MEDIA PREPARATION
FOR THE
BIOLOGICAL SCIENCES

METHODS OF
MEDIA PREPARATION
FOR THE
BIOLOGICAL SCIENCES

By

JOYCE A. STEWART, A.B.

Principal Laboratory Assistant
University of Tennessee, Knoxville

CHARLES C THOMAS • PUBLISHER
Springfield • Illinois • U.S.A.

Published and Distributed Throughout the World by
CHARLES C THOMAS · PUBLISHER
BANNERSTONE HOUSE
301-327 East Lawrence Avenue, Springfield, Illinois, U.S.A.

© *1974, by* CHARLES C THOMAS · PUBLISHER
ISBN 0-398-02990-3
Library of Congress Catalog Card Number: 73 13872

With THOMAS BOOKS *careful attention is given to all details of manufacturing and design. It is the Publisher's desire to present books that are satisfactory as to their physical qualities and artistic possibilities and appropriate for their particular use.* THOMAS BOOKS *will be true to those laws of quality that assure a good name and good will.*

Printed in the United States of America
Y-2

Library of Congress Cataloging in Publication Data

Stewart, Joyce A.
   Methods of media preparation for the biological sciences.

   1. Cultures and culture media. 2. Agar.
I. Title. [DNLM: 1. Culture media—Laboratory manuals. QW25 S8492m 1974]
QR66.S68   1974   576′ .028   73-13872
ISBN 0-398-02990-3

PREFACE

THE PREPARATION OF MEDIA by academic and clinical laboratory
workers is not as simple as the directions or formulas sound.
The techniques and methods used will differ in each laboratory depending on the equipment available; and the knowledge and train-
ing of the "head master." To date there is not an over-all reference
to guide the way. Because of the factitiousness of some of the media,
and because one has to keep card files or notes (which never can be
found when wanted) of formulas for reagents and stains most frequently used, I decided to record a reference manual including the
above. I would also include how-to methods and any ideas which
would prevent the pitfalls and errors I see happen over and over
when inexperienced students are suddenly faced with the task of
preparing media.

I have had former graduate students who have obtained positions,
with Hospitals, State Health Departments, Professorships, School
Teachers call and ask about techniques or methods of preparing
various media. I in turn have called some of them to find out how
to alleviate certain time consuming procedure or better ways to
avoid problems some media present. So the aids, techniques and
methods given here are a composite from various sources and ex-
periences. They are not offered as *absolutes*, but as helpful suggestions.

The methods given in this book will reflect the less expensive
means using ordinary laboratory equipment, rather than the use of
more sophisticated equipment (which is available only when ample
funds have been allotted to the Lab. Prep. Room).

I am deeply indebted to the following people for their patience
in instructing and imparting their knowledge to me: Dr. D. Frank
Holtman, Dr. J. Orvin Mundt, Dr. Stuart Riggsby, and Dr. John
M. Woodward.

I wish to extend special appreciation to the graduate students who gave so much of their time to the project; Mr. Mark L. Camblin, Miss Wanda Graham, Mr. Martin L. Jobe, and Mr. John H. Keene. To Mrs. Becky Birdwell, U.T. Biology Librarian, and Miss Sharon Perkey, Microbiologist, Tennessee State Health Department, a great debt is owed, and to my nephew, D. Stephen Akridge, the Illustrator, gratitude for his thoroughness and patience. For the indulgence and encouragement of my husband, Allen, I am forever grateful.

JOYCE A. STEWART

CONTENTS

METHODS OF
MEDIA PREPARATION
FOR THE
BIOLOGICAL SCIENCES

Chapter I

EQUIPMENT NEEDED FOR PREPARATION ROOM

IT IS ASSUMED that larger equipment such as autoclaves, compressed air lines, dish washers, distilled water facilities, gas burners, gyrotherm, refrigerators, balances and weights, sink with vacuum line or aspirator, sterilization ovens, storage shelves, water baths, work tables and chairs are already in use or available, so the following list is a guide for supplies needed in maintaining a media kitchen for the preparation of media for classroom needs.

Asbestos Gloves

Aseptic Dispensers

Balances

One measuring in grams: one in milligrams

Bottles

Amber bottles and jars. Those which have contained dehydrated media are reusable.

For pouring agar plates in class; 200 ml size with screw caps.

Reagent storage; 1000 ml (six).

Stains; used acid or chemical bottles that have been thoroughly cleaned.

Buckets

For autoclaving used plastic petri dishes; small galvanized scrub tub with bale handle.

3

Bunsen Burners

Burets

Fifty ml size (3-4).

Brushes

Various sizes to clean flasks and test tubes.

Clamps

Burett (Castaloy)*. Designed to hold a buret rigidly vertical when clamped to a support rod.

Cutoff—open (Hoffman). An open end clamp for quick insertion of tubing.

Spring—shut off (Mohr's). All metal construction. Some manufacturers call this type of clamp a flat jawl.

Pinchcock—regular (Castaloy). A spring clip that will close tubing completely without cutting.

Utility—vinylized (Castaloy). Adaptable for a wide variety of holding operations. Length 7": jaw opening 1½".

Clean Up Supplies

Broom, mop, dust pan.

Cotton

Both absorbent and nonabsorbent.

Disinfectant

Mercuric chloride, phenol (5%) or any powerful germicide for general disinfectant can be used for sponging area where plates are to be poured. Dilute as directed.

Dispensing Equipment

See Section Non-Sterile dispensers.

Flasks

Erlenmeyer

4000 ml capacity; (6)

2000 to 3000 capacity; (6)

1500 ml capacity; (12)

1000 ml capacity; (24) for pouring as many as 500-800 plates.

125, 250, 500 ml capacity; sufficient number for classroom needs.

*Scientific Products, 1210 Leon Place, Evanston, Illinois.

Thick wall; Filtering with side arm for connection to filter pump or aspirator.
1000 ml capacity (2)
500 ml capacity (2)

Files
Triangular; for cutting glass.

Filters
Seitz or Millipore®.

Filter Papers
Sizes most frequently used; # 1-7.0 cm, # 50-12.5 cm.

First Aid Kit
Forceps
One; straight tip for removing cover glasses and another one of 8″ or 11″ size, stainless steel, straight blunt points.

Funnels
One long stem and one short stem "58" Kimax for use in filtering solutions.
Three long-stem and three short-stem funnels in sizes 25 to 90 mm diameter for general use; made of glass.
Two powder funnels; top size 65 mm stem diameter 15 mm.

Gauze
Wire gauze squares. 6″ x 6″ and 8″ x 8″ for use on stands over bunsen burners.

Glass Rods
Five to six mm diameter; cut in lengths to use as stirrers.

Glass Tubing
Assorted sizes, 3 mm to 12 mm OD - pyrex.

Graduated Cylinders
One each 10, 25, 50, 100, 250, 500, and 1000 ml.
One—1000 ml plastic is a necessity.

Gyrotherm or Magnetic Stirrer
Magnetic stirring rods; ⅞″ (1); 1⅝″ (2); 2″ (1).
Stirring bar retriever.

Laboratory Coats or Aprons
Metal Trays or Pans
Of sufficient size to hold used pipettes covered with disinfectant.

Paper
Brown Kraft paper—24"-30" for wrapping materials to be sterilized.
Scrap paper—cut in sizes to fit baskets of media for labels.
pH papers.
Weighing papers—6" x 6" size.

Pencils and Pens
Use either lead pencil or markers with permanent ink for labels that are to be used in the autoclave; others smear.

Petri Dishes
Plastic or glass.

Petri Dish Holders or Wire Baskets
Pipette Cans or Bags for Sterilization*
Size 2½" x 1½" x 21"

Rubber Bands
Rubber Stoppers
Rubber Tubing
For gas burners—black gum.
For glass connections—amber gum.

Ringstands
24" height with support stand.

Ring Supports
3" and 5" diameter, inside dimension.

Soap
Hand soap and nonsudsing for dishwasher.

Spatulas
For dehydrated media, one-half dozen four-inch blade length.
For reagents and chemicals, porcelain with spoon on one end and spatula on the other end.

*Propper Manufacturing Co., Long Island City, New York.

Scissors

Large size for cutting paper, cardboard and gauze.

Sponges

For general cleaning and disinfecting plate pouring area.

Spoons

One set household kitchen measuring spoons.

Thermometers

For water bath, range 20° to 100° C with reference lines 37° and 56° C.

Timer

Interval timer of 60 minutes either hand wound or electric.

Test Tubes and Baskets

All sizes necessary for class use.

Wire Baskets

Large size with division for plate storage and other general use. Aluminum baskets for tubes are more serviceable than others for they do not rust.

NON-STERILE DISPENSER, APPROXIMATE MEASUREMENT

A very inexpensive device for dispensing non-sterile solutions into tubes can be made on a ringstand using a 500 ml funnel to which a piece of rubber tubing (about four to six inches long) has been attached. Insert a four-inch part of a 10 ml pipette (the mouth piece) into the bottom of the tubing. Insert clamp on to tubing, pour media in funnel and use clamp to control the amount of media flowing into tubes held in opposite hand. A sample tube with the correct amount of media needed may be held with the tubes in order to obtain an approximate measurement.

NON-STERILE DISPENSER, ACCURATE MEASUREMENT

For a more accurate measurement, another inexpensive device can be made of equipment usually found around a laboratory. Attach a 500 ml funnel near the top of a 36 inch rectangular base support. Directly beside it, using an extension clamp, attach a 50 ml buret

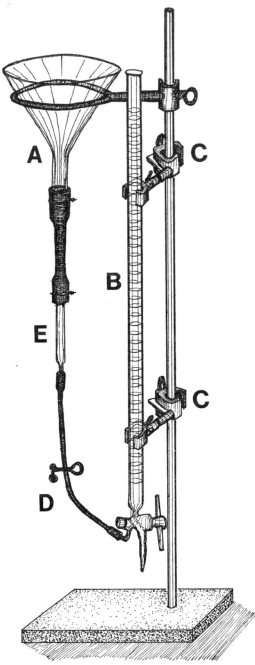

Figure 1. Accurate Measurement Non-sterile Dispenser. A. Funnel. B. Buret. C. Extension Clamps. D. Tubing and Clamp. E. Pipette, rubber tubing wired to funnel and pipette.

(graduated in 1/10's) that has a side arm opening. Use a clamp holder in which an extension clamp has been inserted to do this. Connect the side arm opening and the bottom of the funnel with rubber tubing, on which a spring shut-off clamp has been inserted. Since the buret armopening is smaller than the end of the funnel, it is necessary to graduate the tubing with glass inserts. Again, tip ends from broken 10 ml pipettes serve for this need. Wire all tubing around the glass. If the buret is the open-end type and does not have a cock stopper opening, use rubber tubing and attach a medicine dropper. A spring shut-off or a flat jawl clamp can be used on the tubing to regulate the flow of material.

Other equipment needed for non-sterile dispensing of media is: test tubes and baskets, plugs (either plastic or cotton) and paper or cardboard for labels.

Develop the habit of labeling each basket of media as soon as it is prepared, particularly when several kinds of media such as carbohydrates are being prepared at the same time. Use a lead pencil or felt tip pen with permanent ink when marking labels. Do not use a ball point pen for marking labels for the ink will run in the autoclave.

STERILE ASEPTIC DISPENSER

To dispense media aseptically, all the equipment to be used must be sterilized *in advance*, including preplugged sterile tubes. To make an aseptic dispenser, use a 500 ml leveling bulb with an aseptic filling funnel joined together with rubber tubing. Insert a spring or flat jawl clamp on the rubber tubing between the two bulbs before securing them to the glass stems with wire. In preparing for sterilization, plug both the opening in the top and the little stem in the bell of the aseptic filling funnel with cotton.

AUTOMATIC REFILLING SYRINGE

This device is very inexpensive and can be used for either aseptic or non-aseptic dispensing. An automatic pipetting device is one which can continuously deliver a desired amount (from 0.2 ml to 10 ml) with 0.5% accuracy. The amount to be delivered is set with a screw and locknut on a syringe plunger. An accurately calibrated Luer-Lok®* syringe is used. When the volume is dis-

*Pipettor, Cornwall, Continuous (#13-689, Fisher Catalog, 1973).

pensed, a spring pulls back the plunger drawing the material through a rubber tube which is weighted by a metal sinker to hold it in a flask of the material being pipetted. Delivery needles are obtainable but are not necessary.

Although this device is fast and accurate, there are a few necessary things to know about it to get satisfactory results. One important necessity is *to clean after each use, particularly after using salts and agar*. Run distilled water through the complete device several times.

If the syringe should freeze, drop all assembly in hot water and allow to remain for several minutes. Then take apart and clean thoroughly. If the above procedure does not work, place it in cold water for a few minutes. *Do not try to force it apart*. If the solution won't pump through, yet the barrel is moving, unscrew the sinker attachment and loosen *both* the intake and outgo valves.

To prepare for sterilization, wrap *all assembly* in brown paper and autoclave 15 minutes at 15 lbs. pressure (121° C). To wrap, lay all of the device on the brown paper and bend the hose in a loose loop back up toward the handle. Make sure the hose lays in an easy loop with no sharp bends so that it will not fuse together in the autoclave.

AUTOMATIC PIPETTING MACHINES

There are several makes of electrically operated automatic pipetting machines on the market and each have instruction books regarding the operation of the machine. The common denominator for errors and mistakes are the operators who do not read the instruction book and will not take time to adjust the machine as recommended. Please set the machine as suggested for the amounts of liquid that are to be used and do not use an undersized syringe in the wrong cam opening because it will deliver the amount wanted; doing so causes a great strain on the motor.

These machines are excellent for pouring plates. Advanced preparations are proper settings for the syringe holder and sterilizing the syringe. Wrap the plunger and holder *separately* in brown paper, sterilize 15 minutes then *dry* for 10 minutes. If the autoclave does not have a drying cycle use an oven. Let the packages cool to room temperature before using.

If an environmental controlled hood is available, the pipetting machine can be placed in it and only one person is needed for the operation. But if the work is to be done in an open room there is less chance of contamination if an assistant lifts the lids from the dishes and replaces them during the operation.

Chapter II

MEASUREMENTS

MEASUREMENTS COMMONLY USED

Abbreviations

g = gram
mg = milligram
m = milli (1,000th)
ml = milliliter
μg = microgram (1,000,000th)

Equivalents

1 liter = 1,000 ml
1 gram = 1,000 mg
1 cc = 1 ml
1,000 mg = 1 g

Figures Sometimes Misinterpreted

1 g   = 1,000 mg
0.1 g   = 100 mg
0.01 g   = 10 mg
0.001 g   = 1 mg
0.000001 g = 1 μg (microgram)

Percentages

Percentages are sometimes based on a weight-volume basis, and other times volume-volume. When a one percent (or less) measurement is to be used, a weight-volume calculation is made.
1%—1 g per 100 ml or 10 g per liter. Any amount above 1%, the calculation is made on volume-volume basis.

5%—5 g per 100 ml or 50 g per liter. Measure the 50 g of material and place in a volumetric flask. Use a small amount of distilled water to dissolve materials then add more water to bring measurement up to 1000 ml.

Calculations involving the following percentages are often mistakenly made so we keep this posted in our laboratory.

0.2% = 2 g per 1000 ml or liter
0.02% = 200 mg per 1000 ml or liter

SOLUTIONS

To Make a Percent Solution of a Solid:

Weigh the number of grams required for the desired percentage of the substance to be used. Place in a volumetric flask, add just enough water to completely dissolve the substance, then add more water until the desired measurement is obtained.

Example: 100 ml 5% malachite green

Weigh 5 g of malachite green, place in volumetric flask, use enough water to completely dissolve the substance then add more water to bring up to 100 ml volume.

To Make a Saturated Aqueous, Alcohol, or Acid Solution:

Consult *The Handbook of Physics and Chemistry* for the solubility of the substance.

Example: The number of grams needed to make an aqueous saturated solution of sodium chloride is given in the handbook referred to above under the heading, "Physical Constants of Inorganic Compounds." Solubility in grams per 100 ml if cold water is used, 37.5 g; if hot water is used, 39.12 g.

MOLARITY

The unit of weight for converting compounds of salts and acids is a mole (the gram molecular weight of a compound). The molarity (M) is the number of moles of a compound dissolved in one liter of water. To find the molecular weight of a given salt or acid look on the respective bottle containing the compound for the number by M.W. (molecular weight) or F.W. (formula weight, which is used by some companies).[9]

To make the solution, the solute is measured and placed in a 1000 ml volumetric flask. Add only enough water for the solute to go into solution, then add sufficient water to make the final solution one liter.

A simple formula for obtaining the number of grams of solute to be used in making a molar solution is:

$$\text{Molarity} = \frac{\text{grams of solute}}{(\text{liter of solution}) \times (\text{molecular weight of solute})}$$

Example: How many grams of NaCl would be needed to make a liter of 0.15 M NaCl?

$$.15 = \frac{X}{1 \times 58.44}$$

$(58.44) \quad (.15) = X$

$8.766 \text{ g} = X$

Suppose over a period of several weeks it would be necessary to use either a 0.01 M solute of a 0.001 M solute solution for a formula. A stock solution of 1 M solute would save time.

How much of the 1 M solution would be used for 0.01 M solution to 250 ml?

Using the above formula:

$(X \text{ ml}) \quad (1 \text{ M}) = (.01 \text{ M}) \quad (250 \text{ ml})$

$X = 2.5 \text{ ml}$

Place 2.5 ml of 1 M solution in a graduated cylinder and add water to final volume of 250 ml.

For a 0.001 M solution, obviously .25 ml solution would be used.

NORMALITY

One equivalent weight of any acid or base diluted to one liter final volume with water yields a one normal solution or the acid of base. The normality of a solution may be defined as the number of equivalent weights in a liter of that solution. To determine the gram-equivalent weight of a compound divide the gram formula or molecular weight by the number of hydrogen ions per formula.

Example: NaOH (40 g gram formula or molecular weight) has only one hydrogen ion in the formula, therefore, the gram equivalent weight would be the same (40).

H_2SO_4 (98 g mw) has 2 hydrogen ions so the gram equivalent weight is $98 \div 2 = 49$ g.

$Ca(OH_2)$ (74.09 g mw) has 2 hydrogen ions so the equivalent weight is $74 \div 2 = 37$ g.

MEASURING CHEMICALS AND DEHYDRATED MEDIA

All chemicals and dehydrated media are packaged in as pure a form as possible. It is very important to keep them that way. Use a *clean* spatula or measuring device for each individual product measured, and wash each spatula immediately after use so that it will be clean and not become corroded. Place clean weighing papers on the scale, and using a clean measuring device, start measuring a small amount and build up to the desired amount. This habit is to be developed to avoid waste. It is permissible to take the dehydrated media off the *top* of the amount measured and put it back into the bottle, but this is not true for chemicals.

Weighing Chemicals

Place a very small amount of chemical on a clean weighing paper beside the scale and determine the amount to be weighed from that. *Do not place left over amounts back into the original container. Discard!* Learn to be an expert judge of approximate amounts needed. For the majority of formulas, one fourth to one-half teaspoonful is all that is needed or a small amount on the end of a spatula is sufficient. It is fun to see how close you can come to the desired amount without having any materials left over; some chemicals will really be a surprise—so light or so heavy!

Measuring Dehydrated Media

Directions are given on each bottle of dehydrated medium for the amount to be used for a liter. To determine the amount of medium to use for less than a liter, multiply the number of grams per 1000 ml of the medium by desired number of mls to be made.

Example: If the desired amount to be made is 370 ml and the amount of medium to be used is 46 g per 1000 ml, multiply 46 by .370 = 17.02 g.

If the desired amount to be made is above a liter—1370, multiply 1.370 by 46 = 63.02 g.

If the amount of medium is less than one gram per liter, the same rule applies, but mistakes are often made because of carelessness in placing the decimal point in the proper place.

If the amount to be used is .46 per liter, and 1370 ml were needed, multiply 1.370 by .46 = 6.3 g per liter.

HOW TO DETERMINE THE AMOUNT OF MEDIUM FOR A GIVEN NUMBER OF TUBES OR PLATES

Tubes*

Broth tubes

Generally the amounts used for broth tubes are 7 to 8 ml per tube. EXAMPLE: if 330 tubes are needed, multiply 330 by 8 ml which equals 2640 ml. Round off to 2700 ml; this will make allowance for accidents such as slight spillage during dispensing, etc.

Carbohydrate fermentation or Durham tubes

10 ml per tube. The medium is drawn up into the insert vial during sterilization. The vial must be covered with medium for proper use in testing fermentations.

Slants

8 ml per tube for general use. This allows for approximately a three-fourths inch butt. Use 5 ml for a long slant with a short butt.

Stabs

10 ml per tube or to a depth of 8 cm. Any medium for anaerobes should be deep.

Water blanks

Water evaporates during sterilization; for 16 x 150 mm OD size tube always pour one more ml than specified. EXAMPLE: If specification is for 8 ml, dispense 9 ml.

For 99 ml dilution blanks, the amount to be dispensed before sterilization depends on the amount of evaporation that occurs in the particular autoclaves being used. To determine the correct amount to be dispensed, experiment by placing 4 bottles containing 101, 102, 103, and 104 ml each in the autoclave. After sterilization, measure each one. In our laboratory 102 ml dispensed before sterilization yields 99 ml.

Plates

Plastic petri dishes, 150 x 100 mm—minimum 20 ml per plate, poured by hand.

*16 x 150 mm OD Size Tube.

Glass petri dishes, 115 mm—minimum 18 ml per plate, poured by hand.

Exceptions:

Starch plates are required to be thicker so increase the amount 3 to 5 ml per plate.

Seeded plates (See Chapter VII)

Using plastic dishes: 25 ml per plate.

Using glass dishes: 23 ml per plate.

The reason for the difference in minimum amounts for plastic and glass petri dishes is that the plastic dishes offer more resistance to the flow of medium; consequently, we plan for more agar in order to be sure the bottom of the plate will be completely covered. The medium spreads rapidly over the bottom of glass dishes. We have found that given the same amount of agar to be poured, one in plastic dishes and the other in glass, the yield is always higher in glass dishes due to the factor mentioned above.

Suspensions using bases of broth, water or saline solutions will not spread in plastic dishes; use glass.

Yield

Five hundred ml of sterile medium yields from 23 to 25 plates for general laboratory use. Do not use a 2000 ml flask with a liter of medium in it to pour plates. The flask is difficult to handle and there are more chances for contamination. Instead use 1500 ml flask with 1000 ml or two 1000 ml flasks with 500 ml of medium. The maximum amount of medium to use in a 500 ml flask is 250 ml, for a 250 ml flask, 150 of medium.

THE CORRECT SIZE CONTAINER TO USE FOR MIXING MEDIA

Broths

Medium for broths dissolve easily so a container of a size nearest the amount to be made is suitable if a magnetic stirrer is being used. If the medium is to be shaken by hand, use a container half again as large as the amount to be made.

EXAMPLE: For 3000 ml broth, using a stirrer, use a 4000 ml flask. If shaking by hand, use 2000 ml flasks with 1500 ml in each flask. For smaller amounts use smaller flasks.

Agar

Dispensing Into Tubes

All media containing agar must be heated until the agar is completely melted. The best method is by flowing steam in an autoclave or using a water bath, but these methods take a longer period of time than over direct heat. If using direct heat, the medium must be stirred constantly or it will stick to the bottom of the flask and burn. Also, it must be constantly watched for *as soon as the agar reaches the boiling point, it will bubble up and boil over the container.* Immediately remove from the heat *using asbestos gloves,* let simmer down and replace on low heat and let boil until very clear. Test for complete melting by gently swirling the medium against the sides of the flask, if no particles remain on the sides when swirling ceases, the medium is ready for dispensing. This applies to all agar medium. Most agar media have a clear, opaque look when completely dissolved, but some exceptions, Brain Heart Infusion, Sabouraud's sugar agars, E.M.B., etc., which have a cloudy appearance, can be tested for completion by the swirling method. Be sure the agar is *completely dissolved* before dispensing or else the material will not solidify. Invisible particles will inhibit the growth of some bacteria. (Difco Technical Information #0353, Feb. 1973.)

CAUTION! Although we use the direct heat method for melting agar to be dispensed into tubes, a constant emphasis is made to the person doing so about watching for the boiling point and having asbestos gloves beside the unit so there will not be any danger of the person reaching for the container with the boiling medium with paper towels, etc. *An agar burn is much like a grease burn, the penetration is very deep.*

Use a flask at least one or one and a half times greater than the amount to be made. Beakers or open vessels are not recommended for short, wide vessels require longer periods of time for heat penetration than tall, thinner vessels.[13]

EXAMPLE: 1000 ml flask for 500 ml medium
1500 ml flask for 1000 ml medium
4000 ml flask for 2000 or 2500 ml medium (Because of the narrow neck of the 4000 ml flask, and the weight on handling more than 3000 ml medium, it is recommended that no more than

2500 ml be used in direct heat method of melting agar).

Flowing steam in an autoclave being the most satisfactory and recommended method of dissolving agar medium, allow enough time for agar to *completely melt*, and test by the swirling method before dispensing. The time allowance for 1000 to 2000 ml of medium to dissolve in the autoclave on flowing steam is one to one and one half hours.

Dispensing Into Plates

Agar to be used for pouring plates necessitates a particular concern for the proper size container, because it is necessary to have complete control over the flow of material when poured from the flask into the plate. Individual capability would determine the size flask to use. For people who are inexperienced in pouring plates or ones with small hands, the size that seems to work best is a 1000 ml flask containing 500 ml of medium. For the experienced ones and those with large hands a 1500 ml flask containing 1000 ml medium is very practical and less time consuming.

The agar is placed in the proper size flasks, covered with heavy aluminum foil and placed into the autoclave on flowing steam for ten to twenty minutes (depending on the amount of medium in each flask: 500 ml, 10 minutes; 1000 ml, twenty minutes) then sterilized under pressure at 121° C for 15 minutes. Upon removing from the autoclave, get in the habit of testing for complete melting of the medium by gently swirling the flask (do not create a lot of air bubbles). If the particles remain on the sides of the flask, replace in the autoclave and sterilize for 5 to 10 minutes, depending on the amount of media being made.

Once when we were having difficulty with a certain agar medium taking a longer period of time than usual to melt, I asked the Difco Representative why. He explained that agar is from kelp, a kind of sea weed. Difco had found kelp from different areas of the world had different time elements for melting, and that was why they tested agar from new sources before using, for they tried to have a standard scientific product.

MEASUREMENT OF AGAR IN CONTAINERS
FOR CLASS TO POUR PLATES

If a class exercise specifies an individual flask of agar for pouring several plates, a quick method of preparation is to use single measurements for each flask, since the agar in such small amounts will dissolve during sterilization. (See chart below.) Pour the measured medium, using a powder funnel, into the flask. Add the specified amount of water, plug and sterilize. Experience has shown that a set of ordinary kitchen measuring spoons will suffice for the individual measurements. Although the amounts are given in the chart below, first measure the amount on the balance and then pour into the measuring spoon, in order to become familiar with the exact amount needed and to become familiar with the meaning of level and scant teaspoonsful. After some experience, the chart can be referred to for the amount needed.

CHART FOR MEASURING AGAR MEDIA
INTO TWO HUNDRED FIFTY ml FLASKS:

Specification	Amount of Media	Amount Distilled Water
100 ml Nutrient Agar	1 level teaspoon	100 ml
150 ml Nutrient Agar	1¼ scant* teaspoons	150 ml
100 ml Plate Count Agar	1 scant teaspoon	100 ml
150 ml Plate Count Agar	1¼ level teaspoons	150 ml
100 ml Trypticase Soy Agar	1 scant teaspoon	100 ml

When preparing agar medium to be used in class several weeks later, the most satisfactory containers to use are bottles with screw caps† or flasks with screw caps. Dissolve the medium by boiling and pour through a funnel into the bottles. After filling the bottles with the specified amount, *set screw caps with only one full turn for processing in the autoclaves.* After sterilization, when the bottles have cooled in the autoclave enough to be handled, *screw caps down tightly before setting on storage shelves.*

CHANGING SOLIDIFIED AGAR SOLUTION
INTO A LIQUID FORM

The quickest method is placing in the autoclave on flowing steam.

*Scant teaspoon means *not quite coming up to a stated measurement.*
†Narrow-mouth bottles, #5828, Will Scientific, Inc.

If the medium contains any sugars or carbohydrates other than glucose, it is best to use a water bath. In either case do not leave the medium in the excessive heat for a longer period of time than necessary for dissolving because excessive heat breaks down the nutrient values of some substances.

Since waste is a problem in planning for classroom needs, the question is often asked, how many times can an agar medium be remelted for use in experiments. The circumstances determine the answer. We have found nutrient agar remelted as many as four times serves the purpose for our particular exercises in the basic level, but for the advanced exercises such as "the growth experiment" only fresh medium is used. Any medium containing carbohydrates, including glucose, should not be liquidified but one time for such excessive heat destroys the sugar.

In Clinical Laboratories it is suggested only freshly made medium be used.

THE USE OF PIPETTES IN MEASURING
FROM 1 TO 10 ml OF LIQUID

Using a pipette is not like sipping soda through a straw for control of the amount of liquid taken in is most important. The stopping point on the pipette is marked, and care should be taken that no liquids of any kind enter the mouth. Grasp the pipette near the top with the thumb and three fingers, leaving the index finger free to press against the top of the pipette to hold in the liquid. To release the liquid, lift the finger. All the liquid will flow out of a pipette marked TD, to deliver, but the last drop has to be blown out of a pipette marked BO, blow out. There are times when only part of the liquid is to be released and the question is, how is that done? By retaining pressure of the finger on top to admit only a slight bit of air. The ability to do this well comes easily after practice. For the first attempt, the liquid to use is water.

Since pipettes are of different sizes and graduations, it is necessary to choose the one best suited for the purpose for which it is to be used. There are two kinds of serological pipettes that are usually used in the preparation room:

TD: To deliver. The liquid flows out of this type freely. The top is of clear glass.

BO: Blow out. All the liquid does not flow out of this kind, the

last drop has to be blown out. The top has a frosted ring
around it.

Since either type may be used, *notice which one* is being used. It is
also important to watch for *the calibrations on the pipettes*. Some are
calibrated all the way to the tip end, others have a stopping point
about an inch and a half from the tip. The graduations are usually in
either 1/10th or 1/100th's.

Chapter III

TO ADJUST pH OF MEDIA

THE PH OF A SOLUTION is defined as the reciprocal of the negative logarithm of the hydrogen ion concentration. In simple terms, pH indicates the amount of acid or alkali that a solution contains. The pH scale ranges from 0 to 14. Neutral, neither acidic nor alkaline, is pH 7.0. Acidic solutions have pH values lower than 7 and alkaline or basic solutions have pH values above 7.

0 1 2 3 4 5 6 7 8 9 10 11 12 13 14

acid | alkaline or basic

neutral

For accuracy, a pH meter which determines the pH by electrical methods should be used. Because of various makes of pH meters, no instructions are given here. Follow the instructions given for the instrument to the letter and handle carefully.

In testing solutions for classroom use, pHydrogen paper* or others of equal quality will suffice. The papers are similar to litmus paper, but the color ranges are more graduated. In order to test with the papers, tear off about one inch of the paper, hold it in one hand, draw up a few drops of the medium with a pipette and let a drop flow out onto the end of the paper. Immediately compare the color of the medium on the paper with the color chart supplied with the paper. The color near the end of the paper that corresponds with the color on the chart indicates the pH.[12]

To adjust a medium to the correct pH either a one normal solution of sodium hydroxide or a one normal solution of hydrochloric acid is used.

*Made by Micro Essential Laboratory, Brooklyn, New York.

If the pH of a medium is below 7.0 (acid) or at 7.0 (neutral) and it needs to be basic (above 7.0) add 1N NaOH to adjust the pH to the desired basic level.

If the pH of a medium is 7.0 (neutral) or above 7.0 (basic) and it needs to be acidic (below 7.0) then add 1N HCL to adjust the pH to the desired acidic level.

Since a one normal solution is highly concentrated, use only one tenth ml pipette for adjusting 1000 ml. Increase amounts gradually. Check pH noticing the changes from one color degree to another; judgment can be made of the amount of additions to be made. Avoid increasing the volume by adding too much and having to use first one and then the other solution to effect the changes.

Formulas

1 N NaOH (one normal solution of sodium hydroxide): Place 40 g of sodium hydroxide in a volumetric flask; add 1000 ml of distilled water to make up volume.

1 N HCl (one normal solution of hydrochloric acid): Place 93 ml of hydrochloric acid in a volumetric flask; add 907 ml of distilled water to bring measurement up to 1000 ml volume.

The question arises, when is the best time to check the pH of a medium containing agar? The answer will vary between microbiologists. If the formula is being made up and agar is to be added, we check the pH before the agar is added.

The question always arises when preparing carbohydrate or sugar broths using nutrient broth as the base, when is the pH adjusted, before or after adding the sugar? Again opinions differ. Since nutrient broth has a final pH of 6.8 and the final pH of the media containing a sugar should be 7.0, we adjust the base pH first. Tests have shown that different batches of various sugars sometimes will and sometimes will not change the pH to any great degree. We add the indicator before adding the sugar, and by experience in noting the colors, can tell if any major change has taken place.

CAUTION: Culture media below pH 6.0 or above pH 8.0, will undergo a change in pH toward neutrality with heat.*

*J. O. Mundt, Ph.D., Professor of Microbiology, Univ. of Tenn., Knoxville.

Chapter IV

METHODS OF STERILIZATION

DEFINITIONS

"STERILIZATION MAY BE DEFINED as the complete destruction of all living organisms in, or their removal from, materials by means of heat, filtration or other physical or chemical methods."[14]

"The meaning of the terms *sterile* and *sterilization* is absolute. There is no such thing as *practically sterile* or *nearly sterile*. Either a thing is sterile or it is not sterile."[4]

The methods of sterilization used in microbiology are: (1) steam under pressure (moist heat), (2) flowing steam not confined by pressure, (3) dry heat in specially constructed ovens, and (4) filtration.

HOW TO OPERATE AN AUTOCLAVE

Medium for bacteriological studies is sterilized by moist heat (steam) under pressure in a device known as an autoclave. It and pressure cookers used in the home work on the same principle. The doors have safety locks and there are two movements to get the door locked. First, the rods are locked in the outer ring, then the handles are turned to tighten the door against the rubber seal. Located at the bottom of the chamber are valves to control the amount of steam going into the jacket, the amount of steam going into the chamber, and an exhaust valve to let the steam escape from the bottom of the inner chamber. There is a valve connected directly to the steam supply line that controls the amount of pressure desired. It is usually set for fifteen pounds pressure.

At the top is a safety valve which will blow off should the pressure get too high; a gauge which indicates the amount of pressure;

25

a thermometer which indicates the temperature *inside* the chamber, and an exhaust valve. When materials are placed in the autoclave, the steam supply is turned on, the exhaust valve is opened and *left open until all the air has been driven out of the chamber*. This usually takes about 5 to 8 minutes. All the air should be exhausted when the temperature reaches 100 to 105° C. Then the exhaust valve is closed and when the pressure reaches the desired limit, *the time is noted* and counted for the specified time for sterilization. When the specified time has elapsed, turn off the steam supply and wait for the pressure to drop to zero and the temperature to drop below a 100° before opening the door. Forcing the doors open before the proper time causes troubles. Plugs will blow off the tubes and the medium will spill out. The agar medium will clog the drains.

The Placement of Materials in the Autoclave

Media to be sterilized in the autoclave are usually exposed to a pressure of 15 pounds of steam, temperature of 121° C, for 15 minutes. There are instances when more than 15 pounds pressure is needed to maintain 121° C temperature; adjust the incoming steam supply valve to achieve and maintain 121° C temperature for the entire sterilization period. A longer period of time is required if the material to be sterilized is bulky or closely packed in the chamber. "In order to be sterilized, *every part of the material must be heated to the sterilizing temperature for the necessary length of time*."[4] So be sure that the necessary temperature is maintained for the sufficient length of time.

Care of the Autoclave

Periodic checks should be made of the autoclave to ascertain that the temperature remains at the proper level while operating under pressure, for, if the autoclave does not perform to the correct temperatures under pressure, living organisms are not killed and the materials are not sterile. One of the easiest tests for ascertaining the sterilizing capability of an autoclave is using adhesive tape on which the word *sterile* is printed invisibly, taped on a container. If the autoclave is efficient, the word becomes visible. Keep the inside of the autoclave clean. If media is spilled, clean it out immediately. There is a drain hole beneath the removable pan in the bottom of the chamber that has a wire mesh strainer in it to catch broken glass and other debris. Remove it and clean often.

FLOWING STEAM

Sterilization by flowing steam is used when the pH of a solution is above or below 6.0 to 8.0, or when media such as selienite broth are destructible by excessive heat.

Flowing steam is that which is not confined under pressure, but has the temperature of 100° C, the same as boiling water. To obtain such a condition in the autoclave, leave the exhaust valve open. Sometimes, on a manually controlled autoclave, the incoming steam pressure is so high that pressure builds up. In that case, turn the steam supply back or down (just like the heat on a stove).

DRY HEAT STERILIZATION

The dry heat ovens provided in laboratories are different from regular ovens in that they have a double-walled chamber and are constructed to withstand a very high temperature. The ovens are used for the sterilization of test tubes, flasks, Petri dishes and other clean glassware. Neither solutions nor paper can withstand the high temperature required (170° to 180° C for one and one-half hours or longer). If the oven is packed full with Petri dishes, allow three hours for sterilization. For smaller amounts one to one and one-half hours at 170° C is sufficient for sterilization.

FILTRATION

Sterilization by means of a filter is used for those media which are adversely affected by heat. Among those media requiring filter sterilization are ascitic fluids, blood serum, carbohydrates and urea. For carbohydrates for class use, sterilization is by careful autoclaving (Chapter X). The most commonly used filters in the preparation room are the Seitz or Millipore. The operational set up for either kind is the same.

If the Laboratory is equipped with an air-vacuum, it is necessary to have a water trap to keep the water from going back through the pump motor. The filter flask is attached to it by means of rubber tubing which has a clamp inserted on it previous to the attachment. The clamp is left loose until the material has been filtered, then the clamp is screwed down when the air vacuum is to be turned off. This prevents the cotton from being pushed back down into the media. As a double precaution against the cotton being sucked back into the flask when the rubber tubing is removed from the flask, leave a little tip of cotton folded back over the outside of the arm long enough to project beyond the rubber tubing attachment. Grasp the cotton projection with the fingers and hold while removing tubing. (See Chapter V.)

If the Laboratory does not have an air-vacuum, the filtering is accomplished by gravity using water from the sink with an aspirator on the faucet. It is advisable to have a flask for a water-trap.

HOW TO MAKE AND SET UP A WATER-TRAP FLASK FOR FILTERING OPERATIONS

Equipment needed:

1. Flask with side arm.
2. One hole rubber stopper, size to fit the flask used.
3. Glass tubing, about eight inches long, size to fit rubber tubing to be used for connections.
4. Two pieces of rubber tubing of sufficient length to reach from the sink to the water-trap flask, and one to reach from the water-trap flask to the filtering flask.
5. One clamp. (If jawl type, insert on tubing to connect water-trap flask and filtering flask before making the connections).

Assembly:

1. Attach the long piece of tubing to the side arm of the aspirator on the water faucet and connect to the side arm of the water-trap flask.
2. Insert rubber stopper with glass insert and tubing into the top of the water-trap flask and connect tubing to the side arm of the sterile filter flask. *Leave cotton plug in side arm of sterile flask.* Screw the clamp down until ready to use. Be careful in removing the covering from the sterile filter funnels. If sterile filters are to be inserted (some sterilize the filters in the assembly —we do not) leave the wrappings in such a manner that they can be used to shield the flask from air-borne contamination. In a Seitz filter, leave the cotton in the upper section until ready to use. Regarding the Millipore, leave the paper over the funnel until ready to pour the liquid.

Operation of Filtering Flasks Using Gravity Flow:

1. Screw the clamp down that is on tubing between filtering flask and water-trap flask, turn the water on low pressure, pour medium into funnel.

2. Slowly unscrew the clamp and watch that water does not flow into the sterile flask. (If the water pressure is low, danger of water flowing into the medium will be minimal. If the water is turned on full force, the water will *quickly* back up into the flask of medium.)

Media will drip slowly through a Seitz filter, but flows rather rapidly through a Millipore. After all medium is dispersed, *screw the clamp down before turning off the water*. Failure to do so results in the water backing up into the flask of medium.

Operation of Filtering Flasks Using an Air-Vacuum:

1. Attach tubing from water trap flask to an air vacuum outlet.

2. Pour medium into funnel.

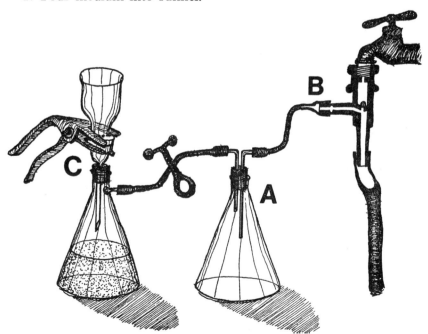

Figure 2. A. Water Trap Flask. B. Aspirator Connection to Sink Faucet and Flask. C. Filter Flask.

3. Turn air-vacuum on slowly until desired rate of filtration is achieved.
4. After all of the medium is dispersed, *screw the clamp on rubber tubing tightly*. Failure to do so will cause the cotton plug in the arm to pop down into the sterile medium.

Chapter V

PREPARATION AND TIMING OF MATERIALS FOR STERILIZATION

ASEPTIC DISPENSERS

BOTH THE BULB TYPE aseptic dispenser and the Luer-Lok are autoclaved intact. Do not set them on the bottom of the autoclave, they will get wet. Set them upon a rack or two tube baskets (anything metal), to keep them off the bottom of the autoclave. Put in autoclave for twenty minutes at 15 pounds pressure (121° C). Exhaust steam immediately and open the door of the autoclave as soon as pressure drops to zero. Let remain in hot autoclave (door open) for 15 or more minutes to dry out.

GLASS PETRI DISHES

Oven Sterilization:

Full oven of closely packed stacks of Petri dishes, time 3 hours at 170° C.

Partially full oven of loosely packed Petri dishes, time 1 to 1½ hours at 170° C.

PIPETTES

Single Pipettes:

Place each pipette in a sterilization bag with the *tip of the pipette down*, the mouthpiece up. Fold top of bag down twice and staple. A handy way to place in autoclave is to use the dish racks from an automatic glassware washer. Stack packages of pipettes one layer deep on each part of the rack then stack racks on top of each other. This allows for better circulation. Place in autoclave for 15 minutes at 15 pounds pressure (121° C). Exhaust steam immediately and open autoclave, referred to as *blow*. If using an automatically controlled autoclave turn to dry vent. Let remain in hot autoclave approximately 15 minutes to dry.

Cans of Pipettes (Wet Sterilization):

Place cans of pipettes in autoclave *with tops off*, setting the tops at an angle by each can. Sterilize 15 minutes as usual. Immediately after opening the autoclave, place the tops on the cans *while in the autoclave* (use asbestos gloves). Do not take cans out of the autoclave to put the tops on, danger of contamination. Transfer to oven for drying; time, 1 hour at 120° C.

Cans of Pipettes (Dry Sterilization):

Place cans in oven with top on. Time, 2 hours at 160° to 170° C.

MILLIPORE FILTERS AND FLASKS

Pyrex Filter Holder (Cat. No. xx 10-047-00): This particular kind is of two parts, a funnel and a base. These are joined together with the filter disc in between by a spring-action holding clamp. The base has a rubber stopper that fits into a standard one liter flask. Both parts and the filter disc have to be sterilized.

1. Insert the base section on the Millipore into a filter flask. Plug side arm of flask with cotton. Wrap both top and side arm with brown paper or heavy duty aluminum foil. We place plastic tape inside the flask over the arm opening to keep cotton from falling into the medium.
2. Use a paper towel or cut paper to fit over the top opening of the funnel, securing with a rubber band. Then wrap *all* with brown paper.
3. Sterile filters are available, but if non-sterile ones are to be used, use forceps (handle carefully; the discs are very brittle and break easily) and place in a Petri dish for sterilization in autoclave.

    CAUTION: Use sterile forceps, either flame or clean with 95% ethyl alcohol.

Hydrosol Stainless Filter Holder (Cat. No. xx 20-047-20): This filter differs from the one previously described in that the funnel locks to the base with the filter disc in between the two. If prefilters are used the whole unit attached to the filter flask can be sterilized at one time. If Millipore filter alone is to be used, wrap separately the funnel and the steel base attached to the filter flask for autoclaving. Plug arm of filter flask with cotton. Sterilize in autoclave for 15 minutes at 15 pounds pressure (121° C). If you take the risk of placing filter disc in the apparatus for sterilization (against the advice of the manufacturer) before using, check to be sure the filter has not broken during sterilization.

SEITZ FILTERS

Assemble with filter disc in place (rough side of disc facing up) on the top of filter flask. Place cotton in top of filter, side arm of flask, wrap all top and side arm opening with brown paper or heavy duty aluminum foil, tie securely with either rubber bands or string, and place in autoclave. Sterilize for 15 minutes at 15 pounds pressure (121° C). Exhaust steam immediately.

SCREW CAP BOTTLES AND TUBES

Place caps on tubes or bottles, but do not screw cap completely down, just enough to hold cap in place. After usual sterilization period open autoclave, and as soon as the tubes or bottles have cooled enough to handle, screw caps down tightly before placing in storage.

WRAPPED MATERIALS

Do not place wrapped materials on the bottom of an autoclave. Instead, set them on something to keep as dry as possible. Be sure each article has enough space around it, so there will be proper sterilization. If autoclave is tightly packed, allow more than the usual fifteen minutes for sterilization time.

STERILIZATION OF CALCIUM CARBONATE TO BE ADDED ASEPTICALLY TO MEDIA

Measure the amount of calcium carbonate and wrap in a small piece of Kraft paper. The paper should only be folded one time to cover the calcium carbonate. Place in a clean Petri dish and sterilize in the autoclave for 20 minutes, 15 pounds pressure (121° C). Add aseptically to media right before pouring plates. Swirl flask until calcium is thoroughly dispersed, and while pouring plates, gently rotate media in the flask.

Chapter VI

POURING PLATES

CORRECT CONDITION OF MEDIUM

ONLY STERILE MEDIA can be used for any purpose in the study of microbiology and, as stated before, sterility is an absolute. Therefore, great care has to be taken when plates are poured so they do not become contaminated. All air and everything a flask or medium might touch carries some bacteria. Sterile medium is to be poured into a sterile Petri dish and neither must come in contact with anything in the process.

CORRECT TEMPERATURE OF MEDIUM

The ideal temperature for medium to be poured into plates is 47° C. However, 47° to 52° C is satisfactory. If a water bath is not available, check the temperature of the medium by holding the flask against the cheek, and if it is not uncomfortably warm, the medium has cooled sufficiently to be poured. If the temperature is too hot, excessive moisture will form when the plates are poured. If the medium is too cool, the medium will be lumpy. Therefore, it pays to wait until the temperature is correct. The penalty for not doing so is *start over*.

The medium may be cooled quickly by rotating the flask slowly under the cold water faucet. But this has pitfalls. If it is shaken too much, air bubbles form in the medium and also appear in the plates. If the flask is not rotated enough while under the running water, the medium gets lumpy. It is not advisable to reheat media.

When a water bath is to be used, turn it on and set the correct temperature before preparing the medium in order to have the bath ready when the medium is taken out of the autoclave. Check the amount of water in it. The water should be sufficient to come up to the top of the media, when flasks are placed in the bath. *Never have the water level below the level of the medium in the flasks.* If several flasks are placed in the water bath, a five inch ring from a tripod is good to put over the neck of flasks to keep them from floating. For those laboratories with ample supply funds, Handi-Wrap®* is most useful.

HOW TO AVOID CONTAMINATION WHEN POURING PLATES IN AN OPEN ROOM

If only a few plates are to be poured, a space in part of the room where there is little air movement will suffice. But if several hundred are to be prepared, it is best to work in a closed room with as little air current as possible. Sponge down the work area including tops of work benches, back stops, gas fixtures (especially the rubber tubing to Bunsen burners) with a phenol or chlorine base disinfectant. Set out the number of plates to be used. For beginners, short rows of single plates are best, but after practice use groups of three. Some people are able to handle stacks of 5 or 6 plates because of the size of their hands—fine!

1, 2, 3 METHOD OF POURING PLATES

1. When the medium has reached the correct temperature take out of water bath and gently swirl the flask for good dispersion. Remove plug or cover from the flask and flame with a Bunsen burner all around the top. Swirl each flask of medium that has been sitting in the water bath in case some have started to solidify; water baths do not remain at constant temperature in rooms where air-conditioning flow is over the water bath. When media is left in the water bath for a long period of time at the temperature for pouring plates (47° to 50° C), the agar will sometimes begin to solidify around the edges and bottom of the flasks, causing sloppy plates with lumps of agar.
2. Open the lid of the plate as little as possible to pour in the medium without touching the flask either to the lid or the plate. Pour in enough medium to cover three-fourths of the plate. *Replace*

*Handi-Wrap, I²R Instruments for Research and Industry, 108 Franklin Avenue, Cheltenham, Pa. 19012.

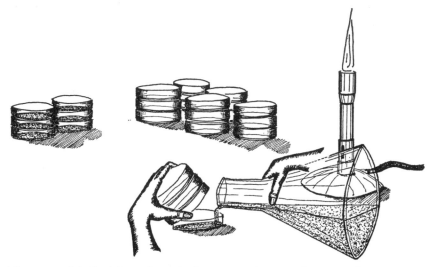

Figure 3. (1) Position of stacked, empty plates to be poured. Area near flame for actual pouring of plate. Note opening of lid. (2) Gentle swirl of filled plate to disperse air bubbles, slide to resting area for hardening.

Figures 3 and 4. 1, 2, 3 Method of Pouring Plates.

Figure 4. (3) Resting flask near flame, while bringing with other hand empty plates to position in pouring area.

lid quickly. This amount will make a plate of sufficient thickness for average classroom needs. Pouring enough medium to completely cover the bottom of the plate will produce a plate thickness of media up to the edge of the lid of the plate.

3. After the three or more plates in one stack have been poured, rest the bottom of the flask of medium (still holding it) beside the Bunsen burner. Meanwhile, gently swirl the filled plates three times one way and three times the other way without getting the medium on the lid. There are two reasons for this procedure: (1) medium does not spread over plastic plates as rapidly as that poured into glass plates, so one is good for dispersion; (2) air bubbles will go to the outside of the plate. *Get in the habit of following the above procedure*. It will become so automatic that little time will be wasted and better plates will be poured. Carefully push the filled plates to the back of the table and bring forward to the flask of medium a stack of empty plates. *Do not take the flask to the empty plates. Bring the plates to the flask*. Flame the neck of the flask after every three or four stacks poured.

HOW TO AVOID CONTAMINATION WHEN POURING PLATES IN AN ENVIRONMENTAL CONTROLLED HOOD

In those hoods which have a glass front there are no problems, if the person uses aseptic techniques in performing the work. But in the open-face ones, two problems arise. First, if not sponged down the accumulated dust swirls around when placing plates and sometimes during the process of pouring the plates. Second, the person pouring the plates must keep from breathing on the plates. Therefore, in our laboratory the workers caution each other to keep their noses out of the hood! If a gas burner is not connected in or near the hood, it is necessary to be extra cautious in handling the flask of medium, to keep from contaminating the rim of the flask. We use aluminum foil to cover about two thirds of the neck of the flask in order to avoid the problem.

HOW TO USE AN ELECTRIC PIPETTING MACHINE FOR POURING PLATES

If the machine has a floor control, one person can use the machine very effectively. Even so, the work goes faster if there are two people to do the work, one to control the flow of medium into the plate and one to open the lid and replace it as soon as the plate is poured.

The greatest hazard is the syringe breaking. If the machine is properly set up as directed, and the syringe is DRY, and near the same temperature as the medium to be used, no problem arises. The syringe must be sterilized, broken down, allowed to dry in the auto-clave, then assembled aseptically before using.

STORAGE OF POURED PLATES

When plates have hardened (usually thirty minutes, depending on the room temperature) *stack upside down* in storage racks or baskets. *Plates cannot be used when there is excessive moisture on the surface of the medium.* By storing upside down, the moisture collects on the lid and not on the medium. Although there is always a slight bit of moisture in the medium, there will not be excessive moisture in the plates, if the medium is of the *correct temperature* before pouring.

Chapter VII

HOW TO MAKE BLOOD AGAR PLATES, CHOCOLATE AGAR PLATES, MILK AGAR PLATES, SEEDED PLATES

A LTHOUGH THE SAME INGREDIENTS are used for both blood and chocolate plates, the differentiation is *temperature*. Regardless of which kind of plates are made, there is a common method of dispersing the blood. Gently agitate the blood to disperse the cells before adding to the medium. Allow the blood to run down the side of the flask as it is delivered from the pipette or syringe. This prevents excessive bubble formation. Pour the blood agar *immediately* into plates. (Do not leave blood agar medium in the water bath for long periods of time.) If air bubbles are present on the surface of the agar in the poured plates remove lid and *quickly* pass the flame of a Bunsen burner over the surface of the agar. Too much heat breaks down the blood cells. Several different kinds of base medium may be used for either blood plates or chocolate plates: Blood Agar base, Trypiticase Soy Agar, or Nutrient Agar containing 0.5 percent sodium chloride.

The base medium for blood agar plates should be cooled to 47° C to 52° C before adding blood which is near the same temperature. If the blood is cooler than the agar, the agar will gel in small lumps. If the base medium is above 47° C to 52° C before the addition of blood, the result is chocolate colored plates only, not good for use

39

as blood plates or chocolate plates. *Throw them out.* If a gyrotherm is available, cool blood straight from the refrigerator may be used. Place magnetic stirrers in each flask of medium *before* sterilization. After the medium has cooled to proper temperature, set on gyrotherm and start stirrer. Distribute blood down the sides of the flask as usual.

To Make Chocolate Plates

Cool the sterilized base medium to *only* 80° C before adding *cold sterile blood.*

Milk Agar Plates

The same procedure as that for blood plates is used for milk agar plates. The temperature of sterile milk to be added to the agar base medium should be 45° C to 47° C. The milk is usually a 5% solution and is autoclaved separately from the base medium for 12 minutes at 15 pounds pressure (121° C).

Seeded Plates

When plates are to be inoculated with bacteria, it is absolutely necessary for the medium to be cooled to 45° C to 47° C. If a controlled water bath is not available, the temperature of the medium may be tested by holding the flask of medium against the cheek for a few seconds without any discomfort. Pour plates immediately after inoculation.

Chapter VIII

HOW TO MAKE SLANTS

THE FIRST REQUISITE for making good slants in baskets is to have the tubes packed straight and tight in a basket with straight sides, no slope. Turn the baskets down, resting them on a 16 x 150 mm OD size tube near the top of the tubes. Always turn the longest side of the basket down, because this makes a more uniform size slant. Check each basket and, if the slant of the tubes in the middle of the basket is too long, use some strong tool (such as a file) and press the bottom of the tubes down in the bottom of the basket. This sometimes makes the top row of tubes too long a butt and too short a slant. Readjust. Handle gently and do not let the media run to the top of the tubes against the plugs. If a basket only partially full is to be made, adjust the tubes in the basket so the bottoms are straight before tilting.

When plastic caps are used the basket method does not work because the tubes cannot be adjusted for even slants. It is necessary to lay the tubes out on some kind of rest piece. We have found rubber tubing to be ideal, and to be practical for making either long or short slants. A greater advantage is the portability of the arrangement.

Racks for slants can be made or purchased ready made.

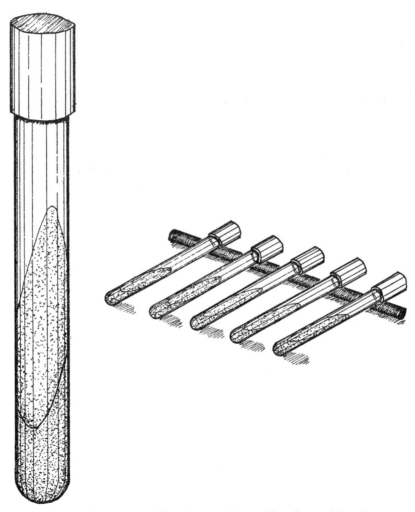

Figure 5. Capped tubes positioned on rubber tubing for making slants.

Chapter IX

HOW TO ROLL COTTON PLUGS

THE ADVANTAGE of rolled cotton plugs is that they can be reused as many as six times, they are faster than plugging with forceps when hundreds of tubes are to be plugged, and they are easier to reinsert in the tubes after inoculation since the ends are firm. If student help is available, they can make them in spare moments while waiting on an autoclave, media to cool, etc. Non-absorbent laboratory cotton* is best to use.

For a test tube size 16 x 150 mm OD, tear off a piece of cotton approximately 8 inches long, 2 inches wide and ¼ inch thick. Begin rolling *tightly*. After rolling ¾ of the plug, tuck in the rough edges on either side and continue to the end. Smooth the finished plug between the forefinger tip and the thumb of the left hand, rotating with the right hand forefinger and thumb. *Do not twist*. If cotton plugs are twisted, open air channels are created and there is likelihood of contamination after sterilization. For tubes smaller or larger than mentioned above, adjust the amount of cotton to be used but use the same procedure for rolling.

All cotton plugs should fit tightly enough in the tube to bear the weight of the tube. They should be long enough to extend into the tube about an inch, leaving enough on the outside of the tube to be grasped between the fingers and held while inoculation is being made. If the plugs are too tight in the tube they cannot be replaced easily. Large plugs for flasks or bottles should be wrapped in cheese cloth or gauze so fuzz from the cotton will not get into the media.

*Rock River Manufacturing Corporation, Janesville, Wisconsin 53545.

Either aluminum or plastic caps may be used for tubes, but the same precaution should be taken in placing the caps on the tubes. The caps should fit tightly enough to withstand the weight of the tube when held by the cap alone.

Chapter X

PECULIARITIES REGARDING DIFFERENT MEDIA

BISMUTH SULPHITE AGAR

Although this precaution applies to all media, it is of particular significance in this instance. *Do not let sunlight shine on plates made of this medium.* A reaction sets in and destroys the usefulness of the medium. Check sterilization procedure given in directions on bottle.[17]

ESCULIN AGAR

When making plates of esculin agar with the addition of ferric chloride or ferric ammonium, do not add the chemicals to the medium before autoclaving. Heat causes the ferric chloride or ammonium to precipitate. Instead add 1 drop into the Petri plate just before pouring medium.

ESCULIN SUGAR

Esculin sugar when added to clear water creates a noticeably bluish cast. Workers sometimes think the flask may have had chemicals in it that did not come out in cleaning and will remake only to find the same result.

MILK

All media made of milk demand careful attention to the sterilization period because overheating during sterilization results in the carmelization of the milk sugar. Again, if this happens, the media *cannot* be used. However, the change in color that takes place when it is autoclaved does not mean the milk has carmelized. It is characteristic that the indicator, either litmus or methylene blue, is reduced to the colorless leuco base. In appearance it will be a yellowish, white color after sterilization. But as the medium cools, it takes on the original color. It is absolutely necessary to remove the medium from the autoclave as soon as possible to prevent overheating, then set in a cool place. If an autoclave is available that regulates temperature and pressure well, set it for 112° to 115° C at 5 pounds pressure for 15 minutes. But if one is in use that is usually kept at 121° C, 15 pounds pressure, sterilize the medium for 12 *minutes only.*

To make litmus milk using dry milk and bacto litmus powder

Dry milk ....100 g

Tap water ....1000 ml

Bacto litmus powder ....0.75 g or 150 mg

Litmus powder or milk solids do not dissolve quickly and easily. Divide the measured water, using 25 to 50 ml for the litmus powder, and the remaining water for the milk solids. Completely dissolve both solutions before adding together.

To make 0.1 percent methylene blue milk

Dry milk ....100 g

Tap water ....1000 ml

Methylene blue ....1 g

Follow the same method as that for litmus milk.

Measurement for dry milk solids

Dry milk ....100 g

Tap water ....1000 ml

NUTRIENT AGAR

The precaution about nutrient agar cannot be repeated too often. *Nutrient agar boils up and over the top of the container as soon as it reaches the boiling point.* Too often, people grab the container without asbestos gloves and, as a result, receive deep burns. Since the medium boils over so quickly, one would assume the agar had dissolved. Not so. It must continue boiling for at least a minute. Remove from the heat. Turn the heat down, reset the medium on low heat and let boil until all agar is melted. It will be opaque. The medium will not solidify if the agar is *not completely* dissolved *before* dispensing. The direct heat method is not recommended, but it is used frequently when time and lack of sufficient number of auto-claves are available.

NUTRIENT GELATIN

The greatest nuisance in making nutrient gelatin is its tendency to gum up in great big clumps, and that which does not clump sticks to the bottom of the flask and burns easily. A gyrotherm is most helpful. If one is available, set the flask of water with the stirrer in it, and, while the water is stirring, pour gelatin in *very, very slowly*. Af-ter complete dispersion, turn on the heat, keep the stirrer going and continue until the medium is dissolved. If a Bunsen Burner is used, heat water to the boiling point and add nutrient gelatin so that moisture from the flask does not hit the gelatin or container it is in. This can be done two ways: (1) make a small cone out of filter paper, the size one can fit in the enclosed hand or, (2) use a stem funnel holding it as far away from the flask as possible. If doing this, use a small funnel and refill frequently. Pour medium in *slowly* and stir vigorously.

If dehydrated nutrient gelatin is not available, use nutrient broth plus 12% gelatin (120 g–1000 ml).

SABOURAUD DEXTROSE, MALTOSE BROTHS, AND AGAR

These media are often overheated because they are always cloudy in appearance even when they are thoroughly dissolved. Therefore, do not let the medium boil over one minute regardless of what it looks like.

S.S. AGAR (SALMONELLA SHIGELLA)

Do not sterilize in the autoclave. There are such few media that are not sterilized in the autoclave that often this fact is overlooked. Read the directions on the bottle. This medium is treated in a very uncommon manner.

SUGAR OR CARBOHYDRATE BROTHS

For most classroom use, 0.5% sugar broths are sufficient. This means 5 g of sugar per 1000 ml. Nutrient broth is the base. Unless otherwise specified, all need an indicator and the most widely used one is Brom Cresol Purple (BCP), 1 ml per 1000 ml.

FOR EXAMPLE: If the need is for 50 tubes each of 4 different kinds of sugar broths, the easiest way to make it would be:

50 tubes $\times$ 4 = 200 total number of tubes needed

10 ml per tube (since inverted gas vials will be used) = 2000 ml

But 2000 ml does not allow for some loss in dispensing, so 2100 nutrient broth base is made with the indicator (2.1 ml BCP) added and then divided into 4 flasks of 525 ml each. Be sure to mark each flask either before or immediately after the particular kind of sugar is added. *Get in the habit of marking sugar broths for one cannot be distinguished from another and costly mistakes can be avoided by developing this habit.* Another reason for making the total amount needed is that the pH should always be checked before adding the sugar and much time is saved by checking one flask.

Overheating causes the sugar to carmelize and the media cannot be used if that happens, *so watch sterilization.* If an autoclave is available that regulates temperature and pressure, set for 112° to 115° C at 5 pounds pressure for 15 minutes. But if one is in use that is usually kept at 121° C, 15 pounds pressure, sterilize 12 minutes only. Take out of the autoclave as soon as possible. Research students refer to references given regarding preparation of carbohydrate broths.[1]

SUGAR SHAKES

Sugar shakes are different from the broths in that they have an agar base, but when they are dissolved the solutions have a similar appearance to the broths. Sometimes workers do not realize the difference and, if not cautioned, will add inverted vials. The method for making is the same as that for broths, but the base medium is nutrient agar. Add sugar *after* the agar has melted, *never before.* Heat breaks down the sugar.

Glycerol agar shakes

Nutrient agar . . . . 21.9 g

Distilled water . . . . 950 ml

Glycerol . . . . 50 ml (measured
in graduated cylinder or beaker)

Brain heart infusion agar 5% glycerol

BHI agar . . . . 49.4 g

Distilled water ....950 ml

Glycerol ....50 ml

When amounts using only 10 ml or less of glycerol is made, it is better to pipette the amount. The pipette can be held over the neck of the flask by a flat-jaw clamp and allowed to drip while the worker is doing something else. Use TD pipette.

THIOGLYCOLLATE BROTH

This medium changes color in processing. As it melts or dissolves it turns red or gold depending on the amount of oxygen introduced into the medium, and it has an obnoxious odor. Also, it takes time to dissolve the solids and agar. The greatest surprise is after autoclaving, it looks like nutrient broth. After cooling, the media oxidizes and the indicator in the upper portion of the tubes turns pink or red. If as much as a third of the tube is of a pink or reddish color, reheat to drive off the absorbed oxygen before using.

MEDIA WITH HIGH CONCENTRATIONS
OF SALT OR SUGAR

There is very little trouble with broth media that have a concentration of 25% to 60% salt or sugar. But with an agar base, difficulties arise when percentages of 25% to 60% are called for. If 25% or more sugar or salt is needed, dissolve the salt or sugar in water over heat, stirring vigorously, and then add the agar base. For concentrations of less than 25%, the salt or sugar and agar can be combined, dispersed as much as possible, and placed in the autoclave for melting and sterilization. But this cannot be done with the higher concentrations. The salt does not dissolve and the sugar sticks in the flask. Dissolve the salt or sugar in water first.

Chapter XI

UTILIZATION OF STANDARD MEDIUM FOR SIMPLIFIED PREPARATION OF A COMPLEX MEDIUM

O UT OF STOCK of a medium, particularly one of many components? Sometimes there is a related one to which you can add the missing ingredient or ingredients to make what is wanted. Two precautions: double check the ingredients as to *kind and amount* of each in both formulas to see if there is a correlation; second, check the final reaction of the two medium, to ascertain if any adjustment in pH is necessary. *Compare and use media of the same manufacturer only.*

NUTRIENT AGAR

Adjust the pH of dehydrated nutrient broth to 7.0, add plain agar. The standard amount of agar to use is 15 g per liter. Remember, check the pH of the broth. The final pH of nutrient broth is 6.8 and that of nutrient agar should be 7.0. One ml of 1 N NaOH per liter usually makes the correction.

NUTRIENT GELATIN

Use nutrient broth base, but do not adjust pH. The final reaction of nutrient gelatin should be the same as that of nutrient broth-6.8. Add 120 g gelatin per liter.

TRIPLE SUGAR IRON AGAR

This medium contains twelve ingredients, eleven of which are in the same amounts as those in Kleiger Iron Agar. If Triple Sugar Iron Agar is unavailable at the time needed but Kleiger Iron Agar is, use K.I.A. Add 10 g saccharose per liter.

MAKING AGAR FROM DEHYDRATED BROTHS

The standard amount of plain agar to add to dehydrated broths to make like-agar medium is 15 g per liter.

These examples are of media used frequently, and are by no means the only ones. A little detective work in the Difco, BBL or other product manuals which are in use in the laboratory will be rewarding. If manuals are not in the laboratory, the companies will supply free copies upon request. The addresses are:

Difco Products
P. O. Box 1058 A
Detroit, Michigan 48232

BBL, Division of BioQuest
P. O. Box 175
Cockeysville, Maryland 21030

Chapter XII

FORMULAS OF MEDIA FOR FROZEN STORAGE OF VARIOUS ORGANISMS

THE MAINTENANCE and preservation of cultures is essential for teaching and research laboratories. Care should be taken in methods and media used. A most comprehensive source of information based on the American Type Culture Collection is found in *Manual of Microbiological Methods*, McGraw-Hill, 1957. One of the newer works, *World Dictionary of Collections of Cultures of Microorganisms* S. M. Martin and V. C. Skerman, New York, Wiley Interscience, 1972, is informative.

The media most frequently used for our classroom laboratories cultures are:

1. Nutrient broth base plus 15% glycerol.

2. Brain Heart Infusion plus 15% glycerol.

For lyophilization (freeze-drying method) of cultures the following medium has been most effective.

Trypticase Soy or Tryptic Broth*	3 g
Yeast Extract	0.1 g
Glucose or Dextrose**	0.2 g
Glycerol or Glycerin	16 ml
Distilled Water	84 ml

Tube 3 ml in one or two dram size bottle. Cap loosely. Autoclave 12 minutes only at 121° C. Remove from the autoclave immediately upon completion of sterilization. Screw caps down tightly.

*Either medium can be used: Trypticase Soy is BBL's product, with Phytone Peptone, and Tryptic Soy is Difco's with Soytone Peptone.

**Glucose and dextrose, as well as glycerol and glycerin are interchangeable words for the same product designated by different manufacturers.

Chapter XIII

PECULIARITIES OF SYNTHETIC MEDIA

THE QUESTION ARISES what is the difference between dehydrated (complex) and synthetic (defined) medium. In dehydrated (complex) medium the chemical composition of all components is not completely defined. In synthetic (defined) media the chemical composition of all known components is defined. With this definition in mind there should not be any question regarding separate sterilization or special handling of the various components, but many workers making the media make the mistake of not following the directions as given for the defined medium and dump everything in without regarding the instructions given. They find out too late, sometimes great precipitation forms or some chemicals crystalize instead of dissolve. To eliminate waste of time and materials, take time to respect the directions given. Such precautions are not usually necessary with complex dehydrated media; these media are prepared by the manufacturer in such a way as to avoid the problems described.

STERILIZATION AND SOLUBILITY OF EIGHTEEN FREQUENTLY USED AMINO ACIDS[16]

L—Alanine—autoclave: Soluble in water.
L—Arginine—autoclave: Soluble in water.
L—Aspartic acid—autoclave: Soluble in water.
                          More soluble in salt solutions; soluble in acids, alkalies, insoluble in alcohol.

L—Cystine—filter sterilize: Quite soluble in aqueous solutions below pH2 or above pH8. Insoluble in alcohol.

L—Glutanic Acid—autoclave: Soluble in water. Insoluble in ether, acetone, cold glacial acetic acid.

L—Glycine—filter sterilize: Soluble in water. 100 g of absolute alcohol dissolves about 0.06 g. Soluble 164 parts pyridine. Almost insoluble in ether.

L—Histidine—filter sterilize: Soluble in water. Slightly soluble in alcohol; insoluble in ether.

L—Isoleucine—filter sterilize: Soluble in water; sparingly in alcohol (hot); insoluble in ether.

L—Leucine—filter sterilize: Soluble in water, less soluble in alcohol; insoluble in ether.

L—Lysine—autoclave: Soluble in water, very slightly in alcohol, insoluble in ether.

L—Methionine—filter sterilize: Soluble in water, warm dilute alcohol; insoluble in absolute alcohol, ether, benzene, acetone.

L—Phenylalanine—autoclave: Soluble in water; slightly soluble in alcohol methanol, ethanol.

L—Prolene—filter sterilize: Soluble in water; very slightly soluble in methanol, ethanol.

L—Serine—filter sterilize: Soluble in water; insoluble absolute alcohol, ether.

L—Threonine—autoclave: Soluble in water; insoluble in absolute alcohol, ether, chloroform.

L—Tryptophan—filter sterilize: Slightly soluble in water—low heat necessary; soluble in hot alcohol; insoluble in chloroform.

L—Tyrosine—autoclave: Soluble in alkaline solutions; insoluble in absolute alcohol, ether, acetone.

L—Valine—filter sterilize: Soluble in water; insoluble cold alcohol, ether, acetone.

Chapter XIV

FORMULAS FOR BUFFERS, INDICATORS, REAGENTS, AND STAINS

BUFFERS

Buffer pH 6.4

Monobasic potassium phosphate	1.63 g
Dibasic sodium phosphate	3.20 g
Distilled water	1000 ml

Dissolve the monobasic potassium phosphate and the dibasic sodium phosphate in distilled water and dilute to 1000 ml.

Dipotassium Phosphate Buffer Solution (0.1M)

Dipotassium phosphate (K_2HPO_4)	22.8 g
Distilled water	1000 ml

Dissolve the dipotassium phosphate in the distilled water.

Monopotassium Phosphate Buffer Solution (0.1 M)

Monopotassium phosphate (KH_2PO_4)	13.6 g
Distilled water	1000 ml

Dissolve the monopotassium phosphate in the distilled water.

Phosphate Buffer (0.5 M)

Sol. A.	Monobasic potassium phosphate	47.6 g
	Distilled water	700 ml

Sol. B. Dibasic potassium phosphate   26.1 g
       Distilled water   300 ml

Adjust to pH 7.6 with further additions of 0.5 M K_2HPO_4 or 0.5 $M–KH_2PO_4$.

Phosphate Buffer (0.1 M pH 7.0)

Monobasic potassium phosphate   6.8 g
Sodium Hydroxide   1.1 g
Distilled water   500 ml

Adjust pH to 7.0 with 2N–NaOH.

Citrate Buffer (0.1 M pH 5.5)

Citric acid   10.5 g
Sodium hydroxide   4.4 g
Distilled water   500 ml

Adjust pH to 5.5 with 2N–Sodium hydroxide.

Dilution Buffer

Dibasic sodium phosphate   7 g
Monobasic potassium phosphate   3 g
Sodium chloride   4 g
Magnesium sulfate $(MgSO_4 \cdot 7H_2O)$ 0.2 g
Distilled water   1000 ml

Dissolve each salt in the order given before adding the next. (This is one of the tricky ones that requires a greater length of time to make; chemicals crystalize if added together before each one dissolves). Autoclave 15 minutes, 15 pounds, 121° C.

Sodium Acetate (.01 M)

Sodium acetate   1.36 g
Distilled water   1000 ml

Adjust pH to 7.0.

To choose a group of buffers is practically impossible. Buffers suitable for either enzymatic or histochemical studies are to be found in the works of George Gomori: *Methods in Enzymology*, Vol. I. New York, 1968.

INDICATORS

Bromcresol Purple (0.4%) or BCP

Bromcresol purple	0.4 g
Ethanol (95%)	500 ml
Distilled water	500 ml

Dissolve the bromcresol purple in the alcohol, add water.

Bromcresol Purple (1.6%)

Bromcresol purple	16 g
Distilled water	500 ml
Ethanol (95%)	500 ml

Dissolve the bromcresol purple in the ethanol, then add the distilled water.

BCP is often stored under the chemical name only: Dibromo-o-cresol sulfonephthalen.

Methyl Red Indicator Solution

Methyl red	0.1 g
Ethanol (95%)	250 ml
Distilled water	250 ml

Dissolve the methyl red in the ethanol. Add the distilled water. Mix well and filter through filter paper.

Phenol Red Indicator Solution

Phenol red	0.2 g
Ethanol (95%)	500 ml
Distilled water	500 ml

Dissolve the phenol red in the ethanol. Add the distilled water and mix. Filter through filter paper.

Bromthymol Blue Indicator

Bromthymol Blue	0.4 g
Ethanol (95%)	500 ml
Distilled water	500 ml

Dissolve the bromthymol blue in alcohol and then dilute with water.

REAGENTS
Acetone-Alcohol

Ethanol (95%)	700 ml
Acetone	300 ml

Mix the two liquids.

Acid-Alcohol

Hydrochloric acid (37%) C.D.	30 ml
Ethanol (95%)	970 ml

Dissolve the hydrochloric acid in the ethanol.

Ammonium Molybdate

Molybdic acid	100 g
Ammonium hydroxide	144 ml
Distilled water	271 ml

Dissolve the molybdic acid in the ammonium hydroxide and distilled water. Slowly, and with constant stirring, pour the solution into 489 ml of concentrated nitric acid and 1148 ml of distilled water. Store in a glass-stoppered bottle.

Barritt's Reagent

Solution A

Alpha-napthol	6 g
Ethanol (95%)	100 ml

Solution B

Potassium hydroxide	16 g
Distilled water to make	100 ml

Store solutions A and B separately.

Benedict's Qualitative Reagent

Solution A

Cupric sulfate crystals	17.3 g
Distilled water	100 ml

Solution B

Sodium carbonate (monohydrate)	117 g
Sodium citrate	173 g
Anhydrous sodium carbonate	100 g
Distilled water	700 ml

Dissolve solution B in the distilled water by warming. Cool to room temperature. Pour solution A into solution B slowly and with constant stirring. Dilute with distilled water to a volume of 1000 ml and mix again.

Biuret Reagent

Solution A

Copper sulfate	2.5 g
Distilled water to make	1000 ml

Solution B

Sodium hydroxide	440 g
Distilled water to make	1000 ml

Store solutions separately in amber-colored bottles.

Cetyl Pyridinium Chloride (0.34%)

Cetyl pyridinium chloride	0.34 g
Distilled water	99.66 ml

Place the cetyl pyridinium chloride in a 100 ml volumetric flask. Dissolve in the distilled water and bring to volume.

Congo Red, Aqueous

Congo red	0.5 g
Distilled water	100 ml
Ethanol (95%)	10 ml

Dissolve the congo red in the distilled water and alcohol.

Copper Sulfate (20%)

Copper Sulfate	20 g
Distilled water	100 ml

Place the copper sulfate in a 100 ml volumetric flask. Dissolve in the distilled water and bring to volume.

Diphenylamine Reagent

Diphenylamine	0.5 g
Distilled water	20 ml
Sulfuric acid	100 ml

Add the sulfuric acid slowly to the distilled water, stirring constantly. Then add the diphenylamine and stir until dissolved.

Ethanol (10%)

Ethanol (95%)   52 ml

Add enough distilled water to make a total volume of 500 ml.

Ethanol (40%)

Ethanol (95%)   210 ml

Add enough distilled water to make a total volume of 500 ml.

Ethanol (70%)

Ethanol (95%)   368.4 ml

Add enough distilled water to make a total volume of 500 ml.

Ferric Chloride

Ferric chloride	12 g
Hydrochloric acid	5.3 ml
Distilled water	94.7 ml

Dissolve the ferric chloride in the distilled water, then slowly pour in the hydrochloric acid.

Hydrochloric Acid (1 N); Alcohol Solution (95%)

Hydrochloric acid (37%) C.P.	83.5 ml
Ethanol (95%)	916.5 ml

Dissolve the hydrochloric acid in the ethanol.

Hydrochloric Acid (0.1 N)

Hydrochloric acid (37%) C.P.   8.4 ml

Dissolve the hydrochloric acid in sufficient distilled water to make 1000 ml total volume.

Kovac's Solution (For Indole Test)

Para-dimethyl-amino benzaldehyde	5 g
Amyl or Butyl alcohol	75 ml
Hydrochloric acid (37%) C.P.	25 ml

Dissolve the para-dimethyl-amino-benzaldehyde in the alcohol. Warm the solution gently using a water bath. After the ingredients are dissolved, carefully add the hydrochloric acid and stir.

Methyl Cellulose Solution (10%)

Methyl cellulose powder	10 g
Tap water	100 ml

Heat the water to 85° C (not boiling) and disperse the methyl cellulose powder in the water. Stir rapidly and constantly while cooling the mixture in an ice bath to about 5° C. The solution is now stable at room temperature and can be stored in a tightly closed container.

Methyl Red Reagent

Methyl red	0.1 g
Ethanol (95%)	300 ml

Dissolve the methyl red in the alcohol and dilute to 500 ml with distilled water.

Alpha-Naphthol Solution

Alpha-Naphthol	5 g
Ethanol (95%)	100 ml

Dissolve the alpha-naphthol in the ethanol. Stir to mix thoroughly.

Alpha-Naphthylamine Solution

Alpha naphthylamine	5 g
Glacial acetic acid	30 ml
Distilled water	75 ml

Dissolve alpha naphthylamine in the glacial acetic acid and pour into the distilled water.

Nitrite Test Reagents

Solution A

Sulfanilic acid	8 g
Acetic acid (5N)	1000 ml
or	
Glacial acetic acid	1000 ml
Distilled water	1000 ml

Slowly add acetic (5 N) or glacial acetic acid to the distilled water, mix thoroughly, add sulfanilic acid.

Solution B

Dimethyl-∞-naphthylamine	6 g
Acetic acid (5N)	1000 ml
or	
Glacial acetic acid	1000 ml
Distilled water	1000 ml

Slowly add acetic acid (5 N) or glacial acetic acid to the distilled water, mix thoroughly, add dimethyl-∞-naphthylamine. Warm in a water bath to completely dissolve.

To make: *acetic acid* (5N) (1 part glacial acetic acid, 2.5 parts water)

Acetic acid	286 ml
Distilled water	714 ml

Add the acid to the water. Mix thoroughly.

To make: *Glacial acetic acid* (99%)

Glacial acetic acid	294 ml
Distilled water	706 ml

Add the acid to the water. Mix thoroughly.

If making up large amounts of solution A & B, for future use, store separately.

Nessler's Reagent

Potassium iodide	50 g in 35 ml distilled water
Mercuric chloride	Saturate solution until a slight precipitate persists
Potassium hydroxide (50% solution)	400 ml

Place potassium iodide solution into volumetric flask, add saturated mercuric solution, add potassium hydroxide, dilute to 1000 ml. Allow to settle, decant the supernatant for use.

Orcinol Reagent

Orcinol	10 g
Ethanol (95%)	100 ml

Place the orcinol in a 100 ml volumetric flask; bring to volume with the ethanol.

Oxalic Acid Solution

Oxalic acid (anhydrous)	20 g

Place in a graduated or volumetric cylinder. Make up to 100 ml with distilled water.

OXIDASE TEST REAGENTS
Para-aminodimethylaniline

Para-aminodimethylaniline	0.1 g
Tap or distilled water	10 ml

Add the dye to the water, let stand for 15 to 20 minutes before using. Do not use until a definite purple color has developed, but use immediately and discard excess. The dye becomes inactive in 1 to 2 hours.

Tetramethyl Para-phenylenediamine Dihydrochloride

Tetramethyl para-phenylenediamine Dihydrochloride	0.1 g
Distilled water	10 ml

Directions same as above.

Phenol 5%

Phenol crystals	50 g
Distilled water	1000 ml

Dissolve the phenol crystals in the water. (CAUTION: Phenol crystals will burn the skin, protect eyes.)

Phosphomolybdic Acid (1%)

Phosphomolybdic acid	5 g
Distilled water	500 ml

Dissolve the phosphomolybdic acid in the distilled water.

Potassium Hydroxide (40%)

Potassium hydroxide	40 g
Distilled water	100 ml

Dissolve the potassium hydroxide in the water.

Potassium Hydroxide Solution (M.R.-V.P. Test)

Potassium hydroxide, C.P. (KOH)	40 g
Creatine	0.3 g
Distilled water	100 ml

Dissolve the potassium hydroxide in 75 ml of distilled water, and allow the solution to cool to room temperature. Then add the creatine and stir to dissolve. Add the remaining distilled water.

Potassium Hydroxide (0.1 M)

Potassium hydroxide (KOH)	5.6 g
Distilled water	1000 ml

Add the potassium hydroxide to 500 ml of distilled water in a volumetric flask. Mix until thoroughly dissolved. Bring to volume with distilled water.

Potassium Phosphate (0.2 M)

Potassium phosphate (KOH)	34.9 g
Distilled water	1000 ml

Dissolve the potassium phosphate in 500 ml of distilled water in a volumetric flask. Mix until thoroughly dissolved. Bring to 1000 ml volume.

Saline Solution (0.9%) Physiologic

Sodium chloride (NaCl)	4.5 g
Distilled water	500 ml

Dissolve the salt in the distilled water. Autoclave at 121° C, 15 pounds pressure for 15 minutes to sterilize.

Saline Physiologic with 10% Methyl Alcohol

To 100 ml methyl alcohol add enough physiologic saline to make 1000 ml.

Silver Nitrate (0.1 M)

Silver nitrate	17 g
Distilled water	1000 ml

Place the silver nitrate in a one liter volumetric flask; bring to volume with the distilled water.

Sodium Chloride (0.85%)

Sodium chloride (NaCl)	8.5 g
Distilled water	1000 ml

Dissolve sodium chloride in the water.

Sodium Hydroxide (1N or 4%)

Sodium hydroxide	40 g
Distilled water	1000 ml

Place the sodium hydroxide in a one liter volumetric flask and dissolve in 500 ml of distilled water; bring to volume.

Sodium Hydroxide (10 N)

Sodium hydroxide	400 g
Distilled water	1000 ml

Place the sodium hydroxide in a one liter volumetric flask; add 700 ml of distilled water. Mix thoroughly. Bring to volume with distilled water.

Sodium Hydroxide (0.1 N)

Sodium hydroxide, C.P. (NaOH)	0.4 g

Place the sodium hydroxide in one thousand volumetric flask, dissolve in 50 ml distilled water. Bring up to volume.

Sodium Hydroxide (0.05 N)

Sodium hydroxide 1N solution	50 ml

Bring up to 1000 ml volume with distilled water.

Sodium Hydroxide (0.2 M)

Sodium hydroxide	8 g
Distilled water	1000 ml

Place the sodium hydroxide in a volumetric flask and bring to volume with distilled water.

Sodium Hydroxide (0.04 M)

Sodium hydroxide	1.59 g
Distilled water	1000 ml

In a liter volumetric flask dissolve the sodium hydroxide with distilled water; bring to volume.

Sodium Phosphate, Monobasic (0.1 M)

Sodium phosphate, monobasic (NaH_2PO_4)	12 g

Add the monobasic sodium phosphate to one liter of distilled water. Mix until dissolved.

Sodium Phosphate, Dibasic (0.1 M)

Sodium phosphate, dibasic (Na_2HPO_4)	14.2 g

Add the dibasic sodium phosphate to one liter of distilled water. Mix until dissolved.

Sulfosalicylic Acid (20%)

Sulfosalicylic acid	20 ml
Distilled water	80 ml

Place the sulfosalicylic acid in a one hundred ml volumetric flask; bring to volume with distilled water.

Sulfuric Acid (3 N)

Sulfuric acid	83.4 ml
Distilled water	916.6 ml

Place the water in a one liter volumetric flask; slowly and carefully add the sulfuric acid.

STAINS

Albert's Diphtheria Stain

Toluidine blue	0.15 g
Methyl green	0.20 g
Acetate Acid (Glacial) C.P. (99%)	1 ml
Ethanol (95%)	2 ml
Distilled water	100 ml

Dissolve the toluidine blue and the methyl green in the distilled water. Then add the acetic acid and the ethanol. Mix well.

Bismark Brown

Bismark Brown	1 g
Distilled water	100 ml

Dissolve the Bismark brown in the water.

Carbol Fuchsin (Ziehl-Neilsen's)

	500 ml	*1000 ml*
Solution A		
Basic fuchsin (90% dye content)	0.3 g	3 g
Ethyl alcohol (95%)	10 ml	100 ml
Solution B		
Phenol crystals	5 g	50 g
Distilled water	95 ml	950 ml

Prepare solutions A and B separately and pour solution A into B slowly, meanwhile stirring the solutions vigorously.

Crystal Violet

CAUTION: Crystal violet is a very light-weight powder easily air borne, in fact, sometimes invisible to the naked eye. Particles will cling unseen to spatulas, balances, table tops, etc. If the area and materials are not cleaned immediately after use, crystal violet may appear *unwanted* in the following substances weighed on the balance. The powder stain is impossible to get out of clothing. If hands become stained to such an extent that soap and water do not remove the stain, a wet wad of cotton dipped in 3% alcohol solution will do the trick, or cotton dipped in a paste of regular cleansing powder (Ajax, etc.) can be used to clean the spot areas.

Aqueous Crystal Violet

Crystal violet (85% dye content)	0.05 g to 0.1 g
Distilled water	100 ml

Mix thoroughly.

Crystal Violet 0.5%

Crystal Violet	1.25 g

Dissolve the Crystal Violet in 250 ml of distilled water.

Crystal Violet 1:2,000

Crystal Violet	0.5 g

Dissolve the Crystal Violet in 1000 ml of distilled water.

Crystal Violet 1:10,000

Make up a 1:2,000 solution of Crystal Violet. To four parts distilled water, add one part of the 1:2,000 solution.

Crystal Violet 1:50,000

After making up the 1:10,000 solution of Crystal Violet, add to four parts distilled water one part to the 1:10,000 solution.

Crystal Violet 1:100,000

After making up the 1:50,000 solution of Crystal Violet, add to one part distilled water one part of the 1:50,000 solution.

Crystal Violet - (Hucker's ammonium oxalate)

Solution A	*100 ml*	*1000 ml*
Crystal Violet	2 g	20 g
Ethyl alcohol (95%)	20 ml	200 ml

Solution B

Ammonium oxalate	0.8 g	8 g
Distilled water	80 ml	800 ml

Thoroughness in dissolving the crystal violet in the alcohol, and the ammonium oxalate in the water, is essential before mixing the two solutions.

Crystal Violet (Tyler's)

Crystal Violet	0.10 g
Glacial acetic acid	0.25 ml
Distilled water	100 ml

Dissolve the crystal violet in the acetic acid. Pour into distilled water. Stir thoroughly.

Flagella Stains

Flagella stain is available commercially and leading microbiologists recommend it be used rather than made in the local laboratory.[12]

Leifson's Flagella Stain

$NH_4Al(SO_4)_2 \cdot 12 H_2O$, saturated water solution	20 ml
Tannic acid (20% water solution)	10 ml
Distilled water	10 ml
Ethanol (95%)	15 ml
Basic Fuchsin (saturated solution in 95% ethanol)	3 ml

Reagents must be added in order listed. Solution should be prepared fresh for use.

Leifson's Flagella stain using:

Flagella stain (obtained commercially)	1.7 g
Ethanol (95%)	35 ml
Distilled water	65 ml

Dissolve the flagella stain in the ethanol. Add the distilled water. To completely dissolve the dye, shake the solution often during a ten minute interval. To store: Dispense in sterile bottle. Stable: approximately two weeks if kept tightly closed.

Gray's Mordant for Flagella Stain

Solution A

    Potassium alum, saturated aqueous solution   5 ml

    Tannic acid (20% aqueous solution)   2 ml
      (a few drops of chloroform must be added
      to this if a large quantity is made up)

    Mercuric chloride, saturated aqueous
      solution   2 ml

Solution B

    Basic fuchsin, saturated aqueous solution   0.4 ml

Mix solutions A and B less than 24 hours before using. Both solutions separately may be kept indefinitely, but deteriorate rapidly after mixing.

Grams Iodine

When sediment is present in bottles of Gram's iodine it is not made correctly. Time is the important factor in making Gram's iodine successfully. If a magnetic stirrer is available, stir the iodine crystals and potassium iodide in a small amount of water from one half hour to an hour. If a stirrer is not available, then let the crystals and potassium iodide stand in the water for six to eight hours before diluting. Either make up the solution early in the morning and let stand until the end of the working day or, let stand overnight before diluting with the remainder of the water.

Total amount of distilled water to be used:	*300 ml*	*900 ml*
Iodine crystals	1 g	3 g
Potassium iodide	2 g	6 g
Distilled water for crystals	20 ml	60 ml
Distilled water for diluting	280 ml	840 ml

Lugol's Iodine Solution

Iodine C.P.	50 g
Potassium iodide C.P. (KI)	100 g
Distilled water	1000 ml

Mix the iodine and potassium iodide in a mortar and triturate with a pestle until finely divided. Add distilled water in small portions to wash the contents into a beaker. Add the rest of the distilled water. Stir until completely mixed.

Lactophenol Solution

Phenol crystals	20 g
Lactic acid	20 ml
Glycerol	40 ml
Distilled water	40 ml

Dissolve phenol crystals in the distilled water, add glycerol. Mix thoroughly. Add the lactic acid slowly to the mixture. To keep the solution in a colorless form, store in a brown bottle.

Malachite Green

CAUTION: Malachite Green like crystal violet is a light-weight powder and is easily air borne. Particles will cling unseen to spatulas, balances, table tops, etc. If the area and materials are not cleaned immediately after use, Malachite green may appear *unwanted* in the following substances weighed on the balance. *There is no cleaning off of hands, clothing, etc.—it has to wear off.*

Malachite green	5 g
Distilled water	100 ml

Dissolve the malachite green in the distilled water.

Methylene Blue Staining Solution (Loeffler's)

Methylene blue	0.3 g
Ethyl alcohol (95%)	30 ml
Distilled water	100 ml

Dissolve the methylene blue in the alcohol, add the distilled water.

Methylene Blue (0.5%)

Methylene blue	1.25 g

Dissolve the mehylene blue in 250 ml of distilled water.

Methylene Blue (1:10,000)

Methylene blue	0.01 g

Dissolve the methylene blue in 100 ml of distilled water.

Methyl Green (1%)

Methyl green	1 g

Dissolve the methyl green in 100 ml distilled water.

Nigrosin (Dorner's)

Nigrosin (water soluble)	10 g
Formalin	0.5 ml
Distilled water	100 ml

Dissolve the nigrosin in the distilled water and immerse in boiling water bath for 30 minutes. Meanwhile, filter the formalin twice through double filter paper. When the nigrosin solution is cool, add the formalin. To store: add 0.5 ml of Formaldehyde (40%) as a preservative. Dispense in sterile bottle, stopper tightly.

Safranin 0.5% Solution

Safranin	0.5 g
Distilled water	100 ml

Dissolve the safranin in the distilled water.

Safranin Staining Solution (Gram's)

	100 ml	*1000 ml*
Safranin	0.25 g	2.5 g
Ethyl alcohol (95%)	10 ml	100 ml
Distilled water	100 ml	1000 ml

Dissolve the safranin in alcohol. When completely dissolved pour into the water with constant stirring.

Toluidine Blue (0.1%) in Ethanol (10%)

Toluidine blue	0.5 g

Make up 500 ml of 10% ethanol, then dissolve the toluidine blue in it.

Wright's Stain and Buffer

It is suggested a commercial brand be obtained because of the quality controls necessary to gain a good product. The day by day attention for two weeks is too demanding. Even when stirred on a gyrotherm with magnetic stirrer, it is necessary to stir each day.

Wright's Stain

Wright's stain powder	.3 g
Glycerol (chemically pure)	3 ml
Methyl alcohol (absolute)	97 ml

Place the powder in a dry mortar; grind with a pestle; add the glycerol and grind together thoroughly. Add the methyl alcohol and mix. Allow to stand for about two weeks in a tightly stoppered flask; *shake each day*. Filter and store in tightly stoppered bottles of amber glass.

MOUNTING FLUID FORMULAS

Lactophenol cotton blue fluid

Phenol crystal	20 g
Lactic acid	20 ml
Glycerin	40 ml
Distilled water	20 ml

Dissolve these ingredients by heating gently over a steam bath. Add 0.05 g of cotton blue dye (Poirrier's blue).

Chloral lactophenol

Chloral hydrate	2 parts
Phenol crystals	1 part
Lactic acid	1 part

Dissolve the ingredients by gentle heating over a steam bath.

Sodium hydroxide - glycerin

Glycerin	10 ml
Sodium hydroxide*	20 g
Distilled water	90 ml

Vasper (For providing an anaerobic seal, does not dehydrate or shrink)

Petroleum jelly	1 part
Paraffin	1 part

Heat the two together until melted. Do not let boil, paraffin is flammable. Mix well by stirring. Pour into flasks.

REFERENCES FOR STAINS AND STAINING TECHNIQUES

The importance of correct stains and staining techniques is not in the scope of this work. Only formulas for stains frequently used have been included. However, the group felt that some mention

*Potassium hydroxide may be substituted for sodium hydroxide.

should be made of references that included methods in order to save time, and in some instances, confusion as to what staining method to use. Particularly as to the Gram stains for smears made from cultures and those made directly from lesions which may be found in Saltys, M.A.: *Bacteria and Fungi Pathogenic to Man and Animals*. Baltimore, Williams & Wilkins, 1963, pp. 495-496.

Other references:

1. Manual of Microbiological Methods by the Society of American Bacteriologists Committee on Bacteriological Technics. New York, McGraw-Hill, 1957.
2. Bailey, W., Robert and Scott, Elvyn G.: *Diagnostic Microbiology*, 3rd ed. St. Louis, Mosby, 1970.
3. Conn, H.J.: Biological Stains, 7th ed. Baltimore, Williams and Wilkins, 1961.

Chapter XV

FORMULAS FOR GLASSWARE CLEANING SOLUTIONS

ACID ALCOHOL 3%

	100 ml	*500 ml*	*1000 ml*
Hydrochloric acid (37%)	3 ml	15 ml	30 ml
Ethyl alcohol (95%)	97 ml	485 ml	970 ml

Add hydrochloric acid to alcohol.

DICHROMATE CLEANING SOLUTIONS FOR GLASSWARE

Nitric Acid

Either potassium dichromate or sodium dichromate can be used. Add 200 g of one of the above to 500 ml of water and dissolve thoroughly. Into this solution add 500 ml of nitric acid.

Sulfuric Acid

A two liter pyrex beaker is necessary for the mixing of this solution, for a glass container breaks when the acid is added. Dissolve 40 g of potassium dichromate and 150 ml water in the beaker. Place container in cold water. Add slowly 230 ml concentrated sulfuric acid. *Sulfuric acid gets hot and bubbles out and burns the skin easily. Protect eyes.*

Chapter XVI

SYMBOLS OF CHEMICAL ELEMENTS AND COMPOUNDS"

REQUESTS FOR MEDIA from the preparation room are often written in chemical symbols only. Some laboratory assistants are unfamiliar with the terms. This section contains the most frequently used chemical symbols in microbiology and are arranged alphabetically for ready reference.

Al	Aluminum
$AlCl_3 \cdot 6H_2O$	Aluminum chloride
Ca	Calcium
$CaCO_3$	Calcium carbonate
$CaCl_2$	Calcium chloride
$Ca(OH)_2$	Calcium hydroxide
$Ca(NO_3)_2$	Calcium nitrate
C	Carbon
Cl	Chlorine
$C_6O_7H_8$	Citric acid
Cu	Copper
Fe	Iron
$FeCl_3 \cdot 6H_2O$	Ferric chloride
$Fe_2(SO_4)_3$	Ferric sulfate
$FeSO_4 \cdot 7H_2O$	Ferrous sulfate
H	Hydrogen
HCl	Hydrochloric acid
H_2SO_4	Sulfuric acid
Hg	Mercury

$HgCl_2$	Mercuric chloride
$HgCl$	Mercurous chloride
I	Iodine
K	Potassium
K_2CO_3	Potassium carbonate
$KClO_3$	Potassium chlorate
KCl	Potassium chloride
$K_2Cr_2O_7$	Potassium dichromate
KOH	Potassium hydroxide
KNO_3	Potassium nitrate
KH_2PO_4	Monopotassium phosphate (mono basic)
K_2HPO_4	Dipotassium phosphate (di basic)
K_2SO_4	Potassium sulfate
Mg	Magnesium
$MgSO_4$	Magnesium sulfate
Mn	Manganese
N	Nitrogen
Na	Sodium
Na_2CO_3	Sodium carbonate
$NaClO_3$	Sodium chlorate
$NaCl$	Sodium chloride
$Na_2Cr_2O_7$	Sodium dichromate
$NaOH$	Sodium hydroxide
$Na_2S_2O_4$	Sodium hyposulfite
$NaNO_3$	Sodium nitrate
Na_2O_2	Sodium peroxide
NaH_2PO_4	Sodium phosphate (mono basic)
Na_2HPO_4	Sodium phosphate (di basic)
Na_3PO_4	Sodium phosphate (tri basic)
Na_2SO_4	Sodium sulfate
Na_2SO_3	Sodium sulfite
$Na_2S_2O_3$	Sodium thiosulfate
NH_4	Ammonium

NH_4Cl	Ammonium chloride
$(NH_4)_2HC_6H_5O_7$	Ammonium citrate (di basic)
NH_4OH	Ammonium hydroxide
NH_4NO_3	Ammonium nitrate
$(NH_4)_2C_2O_4H_2O$	Ammonium oxalate
$NH_4H_2PO_4$	Monoammonium phosphate (mono basic)
$(NH_4)_2HPO_4$	Diammonium phosphate (di basic)
$(NH_4)_2SO_4$	Ammonium sulfate
O	Oxygen
P	Phosphorus
S	Sulfur
Si	Silicon
Zn	Zinc
$ZnSO_4 \cdot 7H_2O$	Zinc sulfate

Chapter XVII

FREQUENTLY USED TERMS THAT ARE INTERCHANGEABLE

THE QUESTION ARISES "why are there two different terms meaning the same thing?" Sometimes the difference is just in spelling, the more archaic forms being replaced. Other terms are due to the manufacturers using the term they like best.

FOR EXAMPLE:

Aesculin	Esculin
Dextrose	Glucose
Glycerol	Glycerin
Levulose	Fructose
Phenol	Carbolic acid
Sucrose	Saccharose
Xylene	Xylol (Do not confuse with Xylose, a sugar).

Chapter XVIII

QUALITY CONTROL OF MEDIA

TEN GUIDELINES

1. Use distilled water.
2. Measure ingredients *accurately*.
3. Stir long enough for the ingredients to be evenly distributed and completely dissolved (especially broths; consult previous information on agar).
4. Patiently stir and dissolve *each salt* in synthetic media before adding another.
5. Dispense in thoroughly clean glassware that has had a distilled water rinse.
6. Autoclave at *correct temperature* and remove from the autoclave immediately when sterilization period has been completed (leaving media setting in the hot autoclave while doing something else is damaging, over exposure to heat!)
7. Cool medium to proper temperature before pouring plates.
8. Turn plates over soon after hardening so moisture will not collect on top of the medium.
9. Incubate or leave out in room over night before refrigerating.
10. Store agar medium in a low temperature and low humidity environment.

Many suggestions have been given in these pages, mainly for the purpose of simplifying complicated processes, and to clarify the reason for some procedures to preserve the nutritional values of dehydrated and synthetic medium.

Although quality control is important for the study of organisms, in a clinical laboratory, a person's life may depend on the reports from the laboratory. All of us are dependent on the results of bacterial control in the food processing industry. Even in the hard industries many procedures are dependent on bacterial control. Therefore, those who prepare media have a responsibility to produce a medium that is as near perfect as possible. Because it is so important, methods have now been devised to test the quality of media prepared either in the local laboratory or that purchased ready-made.

A review of some of the most common causes of a poorly prepared medium was published by Barry, A.L. and Feeney, K.L.: Quality in bacteriology through media monitoring. *Amer J Med Tech*, *33*:387, 1967. Barry, A.L. and Fay, G.D.: A review of some common sources of error on the preparation of agar media. *Amer J Med Tech*, *38*:1, 1972.

Laboratory Assistants, have pride in the fact that *you* are the hub in the wheel, and if it were not for your work and the good quality of it, the eutaxy of life would not exist.

Chapter XIX

MISCELLANY

CONVERSION OF TEMPERATURE CENTIGRADE TO FAHRENHEIT[10]

$^{\circ}C = 5/9 \, (^{\circ}F - 32)$

$^{\circ}F = 9/5 \, (^{\circ}C + 32)$

Absolute zero on the Celius (C°) scale is $-273.15C^{\circ}$, on the Fahrenheit scale is -459.67°.

TO ACID PROOF LABORATORY TABLE TOPS

Consult: Frankel, Sam and Reitman, Stanley: *Gradwohl's Clinical Laboratory Methods and Diagnosis.* St. Louis, 1963, Vol. 1, p. 24.

FOR CONCENTRATIONS OF COMMERCIAL REAGENTS IN MOLARITY, NORMALITY AND PERCENTAGE

Consult: Hodgman, Chas. D., Weast, Robert C. and Shelby, Samuel M.: *Handbook of Chemistry and Physics,* 53rd ed. Cleveland Chemical Rubber, 1972-73.

REFERENCES

1. Bailey, W. Robert and Scott, Elvyn G.: *Diagnostic Microbiology*, 3rd ed. St. Louis, Mosby, 1970, p. 334, 362-363.
2. Benson, Harold J.: *Microbiological Applications*, 3rd ed. Dubuque, Brown, 1968, pp. 183-184.
3. Bradshaw, L. Jack: *Laboratory Microbiology*, 2nd ed. Philadelphia, Saunders, 1973, pp. 301-306.
4. Burdon, K.L. and Williams, R.P.: *Microbiology*, 6th ed. New York, Macmillan, 1968, p. 302.
5. Conn, H.J.: *Biological Stains*, 8th ed. Baltimore, Williams and Wilkins, 1969.
6. *Difco Manual of Dehydrated Culture Media and Reagents for Microbiological and Clinical Laboratory Procedures*: 9th ed. Detroit, Difco Laboratories, 1967.
7. Goromi, George: Preparation for buffers for use in enzyme studies, In Colowick, Sidney P. and Kaplan, Nathan O. (Eds.): *Methods in Enzymology*. New York, Academic, 1968, Vol. 1.
8. Goss, Robert C.: *Experimental Microbiology Laboratory Guide*, Ames, Iowa S.U., 1967.
9. Holum, John R.: *Principles of Physical, Organic, and Biological Chemtry*. New York, Wiley, 1969.
10. Hodgman, Chas. D., Weast, Robt. C. and Shelby, Samuel M.: *Handbook of Chemistry and Physics*, 53rd ed. Cleveland, Chemical Rubber, 1972-73.
11. Pauling, Linus: *College Chemistry*. San Francisco, Freeman, 1964.
12. Pelczar, M.J., Jr. (Chairman): *Manual of Microbiological Methods*. New York, McGraw-Hill, 1957, pp. 26-27, 87.
13. Ribbons, Norris: *Methods in Microbiology*, 1st ed. New York, Academic, 1970, p. 285.
14. Salle, A.J.: *Fundamental Principles of Bacteriology*, 4th ed. New York, 1964, p. 201.
15. Soltys, M.A.: *Baoteria and Fungi Pathogenic to Man and Animals*. Baltimore, Williams and Wilkens, 1963, pp. 495-496.

16. Stecher, Paul G. (Ed.): *The Merck Index of Chemicals and Drugs,* 8th ed. Rahway, Merck, 1968.

17. Waterworth, Pamela M.: The action of light on culture media. *J Clin Path,* 22:273-277, 1969.

18. White, Emil H.: *Chemical Background for the Biological Sciences.* Englewood Cliffs, Prentice-Hall, 1964, pp. 32-33.

INDEX

NOTES

NOTES

NOTES

NOTES

NOTES

NOTES

NOTES

NOTES

NOTES